19⁹⁵
9)

The End of Books—Or Books without End?

The End of Books—Or Books without End?

Reading Interactive Narratives

J. Yellowlees Douglas

Ann Arbor

The University of Michigan Press

First paperback edition 2001
Copyright © by the University of Michigan 2000
All rights reserved
Published in the United States of America by
The University of Michigan Press
Manufactured in the United States of America
♾ Printed on acid-free paper

2004   2003   2002   2001   5   4   3   2

*A CIP catalog record for this book is available
from the British Library.*

Library of Congress Cataloging-in-Publication Data

Douglas, J. Yellowlees, 1962–
    The end of books—or books without end? : reading interactive
    narratives / J. Yellowlees Douglas.
        p.   cm.
    Includes bibliographical references and index.
    ISBN 0-472-11114-0
    1. Fiction—Technique—Data processing. 2. Fiction—20th
century—History and criticism—Theory, etc. 3. Experimental
fiction—History and criticism—Theory, etc. 4. Authors and
readers—Data processing. 5. Creative writing—Data processing. 6.
Postmodernism (Literature). 7. Narration (Rhetoric). 8. Closure
(Rhetoric). 9. Hypertext systems. 10. Computer games. 11. Literary
form. I. Title.
PN3377 .5.C57   D68   1999
808.3'0285—dc21   99-6689
                                                              CIP

ISBN 0-472-08846-7 (pbk. : alk. paper)

Acknowledgments

In 1986 John McDaid, then a fellow graduate student at New York University, suggested I meet Jay Bolter, who arrived bearing a 1.0 beta copy of Storyspace. When he opened the Storyspace demo document to show McDaid and I a cognitive map of the *Iliad* represented as a hypertext, my fate was clinched in under sixty seconds. I had seen the future, and it consisted of places, paths, links, cognitive maps, and a copy of *afternoon, a story,* which Jay also gave us.

Within a week, I had upgraded my Macintosh, changed disciplines, and begun work on a dissertation proposal using the same 1.0 beta version of Storyspace, the equivalent of volunteering as a crash-test dummy to check the safety of cars during head-on collision.

After weeks of alternately immersing myself in *afternoon* and losing data to system crashes every ten minutes, I admitted I needed some direction and began frantically trying to locate Michael Joyce, listed on the Storyspace startup dialogue box along with Bolter and William Smith as the authors and developers of Storyspace. The third Michael Joyce in the Jackson, Michigan, white pages answered the phone with a deep, resonant voice.

"I'm looking for the Michael Joyce who wrote *afternoon,*" I said quickly.

"Yes, but you can't know that," he said, first, not realizing I had a copy of *afternoon,* then: "And who are *you?*"

The rest, as they say, is history—and also the beginning of this book.

Even the most haphazard reader of this book will recognize that, without Jay Bolter, Michael Joyce, and Stuart Moulthrop, my work would not exist. Their ideas, interactive texts, critical writings, and work on interface design have long provided me with both rich fodder for my research and an abundance of ideas, methods, and critical and theoretical approaches to interactivity in general. Together they have inspired me in every sense of the word.

To Gordon Pradl I also owe several debts: for his insights on the distinctions between "inner-" versus "other-directed," for introductions to many of the sources that have informed my entire outlook on reading and interactivity, and for giving me the right shove at exactly the right time.

Contents

An Interactive Narrative Timeline

1759 Laurence Sterne's *The Life and Opinions of Tristram Shandy* first appears in print.

1776 Samuel Johnson declares, "Nothing odd will do long. *Tristram Shandy* did not last."

1914 James Joyce's *Ulysses* first appears in print.

1915 Ford Madox Ford publishes *The Good Soldier: A Tale of Passion.*

1938 Louise Rosenblatt's *Literature as Exploration* argues for the importance of the transactions between readers and texts.

1939 Roosevelt's science adviser, Vannevar Bush, describes a hypertext-like device, the Memex, in "Mechanization and the Record."

1949 Jean Paul Sartre's *What Is Literature?* introduces readers to the central tenets of what later becomes known as reader-response theory or reader-centered criticism.

1960 Marc Saporta publishes *Composition No. 1,* a novel on cards.

1962 Douglas Engelbart's paper "Augmenting Human Intellect: A Conceptual Framework" describes the document libraries, multiple windows, and links between texts that later become parts of the AUGMENT system.

1963 Julio Cortázar's *Rayuela* (Hopscotch) includes alternative orders for reading its segmented text.

1965 Ted Nelson coins the term *hypertext.*

1967 The *Atlantic Monthly* prints John Barth's "The Literature of Exhaustion," which suggests that "intellectual and literary history . . . has pretty well exhausted the possibilities of novelty."

1990 Michael Joyce publishes *afternoon, a story,* the first hypertext novel.

1992 Robert Coover introduces mainstream readers to the possibilities of hypertext fiction in the provocatively titled essay "The End of Books," in the *New York Times Review of Books.*

1997 W. W. Norton Company includes two hypertext fictions in its *Postmodern American Fiction: A Norton Anthology* alongside works by Barth, Burroughs, Gass, and Pynchon.

1998 Flamingo, a British imprint of HarperCollins, publishes Geoff Ryman's *253: The Print Remix,* a print version of the hypertext narrative *253,* first published two years earlier on the World Wide Web.

Introduction: The Book Is Dead,
Long Live the Book!

In early 1998, some dedicated bibliophiles were discovering they could download the Jane Austen novel *Emma* from the World Wide Web to their printers at no cost—if you did not count the price of a ream of paper and a couple of hours of local telephone charges and connect time, which amounted to slightly more than the price of a softcover compendium edition of Jane Austen's works. At the same time, Flamingo, a British imprint of HarperCollins, published science fiction writer Geoff Ryman's *253: The Print Remix*, a paperback version of a hypertext narrative anyone could view on the World Wide Web, theoretically, for free.

Less than six months before Ryman's hypertext novel appeared in print, W. W. Norton, the publisher whose anthologies have created the fodder for English literature courses and curricula across America, rolled out its latest anthology, *Postmodern American Fiction*. Its table of contents listed innovators in American fiction between 1945 and 1997, including William S. Burroughs, Norman Mailer, Thomas Pynchon, Don DeLillo, Donald Barthelme, and the authors of two hypertext narratives, Michael Joyce, author of the first hypertext novel, *afternoon*, and J. Yellowlees Douglas, author of the hypertext short story "I Have Said Nothing." On the book's inside back cover the publishers had also affixed a white sticker listing the URL (Universal Resource Locator) of the anthology's accompanying website and a single-user password, enabling readers to sample Web-based versions of the two hypertexts, also represented in brief printed extracts within the anthology itself.

HarperCollins and W. W. Norton have acknowledged hypertext fiction, ironically, by trotting out versions of it in print—exciting a brief flurry of critical applause and a few tut-tuts in online companions to print periodicals including *Atlantic Unbound*, *AP Online*, and *The Cybertimes*, while earning a jeremiad in the *New York Times Book Review* from, all of things, the editor of a Web-based magazine named for the literary salons once frequented by the giants of nine-teenth- and early-twentieth-century print fiction. Welcome to the world of interactive narratives: the Book is dead; long live the Book.

A mere fourteen years ago, Norman Holland and Anthony Niesz noted that "interactive fiction is mostly a fiction"—great concept, a shame about the demo models.[1] Today, we have hypertext novels, novellas, and short stories on disk, CD-ROM, and the World Wide Web, journals both in print and online publishing special hypertext issues, academic treatises mulling over the possibilities for computers and storytelling, clusters of websites listing interactive narratives and criticism, under-graduate and graduate courses exploring the poetics and aesthetics of interactive narratives. Interactive narratives have been the subject of articles on National Public Radio, as well as in *Spin, Details,* the *New York Times Book Review,* the *New York Times Cybertimes,* and the *Guardian*—Britain's only left-leaning daily—plus no fewer than two documentaries by the BBC. But if interactive narratives seem as omnipresent these days as the World Wide Web, examples of them are still nearly as thin on the ground, proportionately speaking, as they were ten years ago, when Internet use was mostly still restricted to academics, hard-core researchers, techies, and geeks.

Some of this avalanche of attention comes from interactive narra-tives representing something distinctly new in an era when newness, in fashion, in film, and in our cultural mainstream, is restricted mostly to revived artifacts that have been sitting out the past few decades: bell-bottoms and platforms, Volkswagen Beetles, hip-hop retreads of Herb Alpert and Joni Mitchell, cinematic remakes of *Mission Impossible, The Saint,* and *The Avengers,* postmodern pastiches, borrowings accompanied with a wink and a nudge. Some critics are excited at the possibilities of harnessing fiction to the power of a com-puter, realizing, perhaps, a reader-centered text, elevating the reader's historically humble role to something approximating the creative energies of its author. Others are equally exercised at the sacrilege potential in, say, yoking the art of fiction once practiced by the cast of authors straight out of F. R. Leavis's *The Great Tradition* to an appara-tus in which Bill Gates is a stakeholder.

If you survey all the print, both tangible and online, that has been dedicated to interactive narratives, however, both enthusiast and Lud-dite camps share two features in common. Only a handful of critics on either side have read more than one or two examples of works in the new medium. Fewer still appear to have performed what could pass for a close reading on even a single interactive narrative before arriving at pronouncements about the medium's value and future potential—a bit like deciding cinema could never yield up a work of the sophistication of *Citizen Kane* or *The Godfather* after watching Fred Ott's nickelodeon-era short *The Sneeze*. And, for all the critical hand-waving and hand-wringing about interactive narratives being reader-centered, fewer than a dozen articles have ever deigned to give the experiences of reading them so much as a cursory glance.[2] Probe more deeply into the field, and you will discover that remarkably little consensus exists as to the definition of interactive narratives as a genre—or even if such a thing exists—let alone the definitions of what constitutes "hypertext," "interactive," and, even, "narrative."

Much of the literature dedicated to codifying, evaluating, and criticizing interactive narratives tends to sift the wheat from the chaff entirely on the basis of a purely imaginary ideal. Even when addressing the experience of reading actual interactive narratives, critics seem to focus less on the texts than on their own treasured memories of reading, fond recollections of live-action game-playing, or vague notions of what a marriage of digital fluidity and narrative fiction ought, ideally, to achieve. Sven Birkerts in *The Gutenberg Elegies* and Laura Miller in the *New York Times Book Review* admit to—as Miller puts it—"meandering" through one and three works, respectively. Miller briefly describes an amble through *afternoon, Victory Garden,* and Mark Amerika's Web-based *Grammatron,* while Birkerts bases his assessment of the medium on what appears to be little more than an hour spent with a single text. More bizarrely, pieces in recent scholarly journals have attempted to survey the strengths and weaknesses of the field with nary a mention of a specific writer, text, or concrete example. In "Poles in Your Face" in the *Mississippi Review,* the only text author Jurgen Fauth mentions by name is Klaus Kieslowski's film *Blind Chance,* a linear narrative that features three distinctly different plot outcomes, while his references to hypertext fiction are limited entirely to citations from brief World Wide Web articles that mostly misquote hypertext theorists Joyce and Jay Bolter.[3] A review in *Postmodern Culture* of hypertext fiction is confined to a deadpan semiotic reading of publisher Eastgate's catalog cover and the contents listed in it—a bit like a *New York Times* reviewer recommending or condemning a novel solely on the strength of the design on the dust cover, or the look of the author's photograph on the inside back flap.[4] Kids, please, don't try this at home.

Just How Interactive Is Interactive?

Even critics who have conscientiously read the works about which they write, like Espen Aarseth in *Cybertext: Perspectives on Ergodic Literature* (1997) and Janet Murray in *Hamlet on the Holodeck: The Future of Narrative in Cyberspace* (1997), do not exactly see eye to eye, arriving, for example, at wildly different interpretations of the term *interactive*. Guided by the definition supplied by Andy Lippman of the Media Lab at the Massachusetts Institute of Technology (for a detailed description of this, see chapter 3), Aarseth insists that what some critics call interaction "is perhaps better described as participation, play, or even use. It is not an apt description of a work where the user can contribute discursive elements to the effect that the 'theme' of the 'discourse itself' is unknown in advance or subject to change."[5]

While Aarseth's definition of interactive does not shove interactive narratives off the board entirely, he does rank it hierarchically below what he calls "cybertext . . . a machine for the production of a variety of expression."[6] In cybertext, the actual content of the text may be determined by a script that enables the computer to evolve its own stories, as in Talespin, a program that generated fables according to parameters written into the application, or Racter, a program that solicited user input with a Rogerian model for interaction as fodder for vaguely surreal or downright dadaesque poetry.[7] Other forms of cybertext include users working within in a Multi-User Dimension (more commonly known by its acronym, MUD) or an Object-Oriented MUD (usually referred to as a MOO) to create both content and formal shapes of a text, or games like *Adventure*, where the content of the grail quest requires the players to tussle with a requisite number of trolls or labyrinthine caves but does not require that the player confront all of them or even a prescribed number and sequence of trolls, gates, or caves. Cybertext, Aarseth argues, is dynamic, whereas hypertext is static: both the content of the hypertext and the permutations of it potentially open to the reader are fixed.[8] In some readings of "I Have Said Nothing," for example, the narrator's brother may drive to the spot where his girlfriend was killed and step in front of a speeding car. In others, he may be restrained by his mother, standing at the curb—or may not even revisit the scene of his girlfriend's car accident at all. Yet the reader is merely following links already constructed by the author, realizing several of the author's scripted permutations of the narrative, Aarseth argues, even though he acknowledges, "When a system is sufficiently complex, it will, by intention, fault, or coincidence, inevitably produce results that could not be predicted even by the system designer."⁹ Thickets of links for each segment of text and specific conditions under which readers activate them can create a hypertext capable of generating loops or narrative permutations even its author never imagined. While the number of narrative versions may be fixed in theory—since content, links, and linking conditions are set by the author—the experience of reading it may, however, feel dynamic, as few readers return to any hypertext sufficiently persistently to exhaust all its possible iterations. If introducing the computer into the formerly cozy Author-Text-Reader triangle creates an Author-Computer-Text-Reader rectangle, Aarseth is plainly more interested in the Author-Computer-Text triangle nestled within it.

Murray, on the other hand, steers mostly clear of the term *interactive* and plumps for *participatory*, texts in which "we can induce the behavior. They are responsive to our input. . . . This is what is most often meant when we say that computers are interactive."¹⁰ In Murray's view, the text exists less as an apparatus to produce a collaboration of human and machine than as a conduit for an immersive, aesthetic experience that invites readers' participation.¹¹ While she is wary of genres and texts that limit readers' sense of agency—evident in her critique of genres like participatory dinner theater that merely situate participants within preexisting scripts—unlike Aarseth, she is not concerned with the cogs grinding away beneath the surface of the narrative. Instead, she values the reader's sense of participation in the unfolding of a narrative and the impact of participation on our experience of art: graphic, auditory, word-based. "Agency," she notes, "is the satisfying power to take meaningful action and see the results of our decisions and choices" (126), an emphasis that gives her survey of what she calls "digital narrative" (51) the scope to consider everything from Joseph Weitzenbaum's *Eliza*, the digital Rogerian therapist, to disk-based hypertext fiction like *afternoon* and the popular CD-ROM game *Myst*. Murray's notion of digital narrative is also congruent with Jay Bolter's definition of interactive fiction in *Writing Space: The Computer, Hypertext, and the History of Writing* as electronic writing containing episodes or topics, connected by decision points or links.¹² For Bolter, the primitive adventure games circulating during the seventies and eighties were themselves a type of narrative, examples of what he called "the nickelodeon era of interactive fiction."¹³

For our purposes, we will rely on the understanding shared by Bolter and Murray and consider *interactive* texts to be those that contain episodes in the form of chunked text and a range of action accompanying a single decision—as in a player's decision to make his or her way to the port side of the sinking *Titanic,* in hopes of getting off the doomed ship in the CD-ROM *Titanic: Adventure Out of Time.* Unlike the pages of a book, however, these episodes are joined together by links, which may be available to readers as choices for navigating through the narrative, like the map of the United States and interstate routes in Matthew Miller's World Wide Web fiction "Trip." Links may also be invisible, triggered when players click on suggestive words in *afternoon* or decide to sweet-talk the deskbot in *Douglas Adams' Starship Titanic.*[14] The experience of reading Michael Joyce's rich prose in his hypertext fiction written for the World Wide Web, "Twelve Blue," however, is radically different from maneuvering one's way through the characters and Orient Express cars in Jordan Mechner's *The Last Express.* To distinguish between different kinds of interactive narratives, we will call text-based narratives like "Twelve Blue" and Stuart Moulthrop's *Victory Garden* "hypertext fiction" and, following Janet Murray's lead, refer to image-based texts like *The Last Express* and Shannon Gilligan's Multimedia Murder series as "digital narratives."

Hypertext Fiction versus Digital Narrative

Much of the ink, both physical and virtual, spilled over interactivity has been focused on hypertext fiction, in part because its most prominent titles have been created by writers laboring over characters, plots, and prose in much the same way Henry James, James Joyce, and other luminaries of English fiction toiled over their stories and novels—and because the whole notion of readers making tangible decisions to experience works of fiction calls into question the roles of author and reader and, even, exactly why we read for pleasure. Digital narratives, on the other hand, hardly threaten to, as Sven Birkerts puts it in his *Gutenberg Elegies,* eclipse "le mot juste and . . . gradually, the idea of the author as a sovereign maker."[15] Like films, digital narratives are produced by teams; even when well-known writers like Jordan Mechner, author of the best-selling game *Prince of Persia,* create scripts for works like *The Last Express,* the origin of the story and identity of the writer is considered about as essential to the final narrative as Herman Mankiewicz's screenplay was to *Citizen Kane,* a fact known mostly by the bona-fide aficionados. Moreover, digital narratives primarily follow the trajectory of *Adventure,* a work considered venerable only by the techies who first played it in the 1970s, cybergaming geeks, and the writers, theorists, and practitioners who deal with interactivity. Hypertext fiction, on the other hand, follows and furthers the trajectory of hallowed touchstones of print culture, especially the avant-garde novel. In the single article that arguably has made more readers aware of hypertext fiction and inflamed more critics than any other, Robert Coover commented on the relationship between print and hypertext in "The End of Books," published in that bastion of print criticism, the *New York Times Book Review:*

> In the real world nowadays . . . *you will often hear it said* that the print medium is a doomed and outdated technology. . . . Indeed, the very proliferation of books and other print-based media, so prevalent in this forest-harvesting, paper-wasting age, *is held to be a sign* of its feverish moribundity, the last futile gasp of a once vital form before it finally passes away forever, dead as God.
>
> *Which would mean* of course that the novel, too, as we know it, has come to its end. Not that *those* announcing its demise are grieving. For all its passing charm, the traditional novel . . . *is perceived by its would-be executioners* as the virulent carrier of the patriarchal, colonial, canonical, proprietary, hierarchical and authoritarian values of a past that is no longer with us.[16]

A careful reading of the piece reveals just how adroitly Coover can both suggest fiction is ready for its next evolutionary leap and describe the aesthetic potential of hypertext fiction without ever personally and explicitly pronouncing the novel's imminent demise himself—a smart move, coming from an early champion of hypertext fiction who still turns out highly readable novels with a fair degree of regularity. "The End of Books," nonetheless, enraged some critics, a few proving to be literal-minded readers adept at glomming onto the big picture but poor at taking in subtle details. Writing in the *New York Times Book Review* nearly six years later, critic Laura Miller still splutters with palpable rage as she recalls Coover's suggestion: "Instead of following a linear story dictated by the author, the reader could now navigate at will through an 'endless expansion' of words."[17] In "The End of Books," however, Coover perceived that expansion as problematic, a tendency that could turn narrative flow to slurry, making it "run the risk of being so distended and slackly driven as to lose its centripetal force."[18] When Miller tells us that "[p]roclamations about the death of the novel . . . can still get a rise out of a surprising number of people," we know she is speaking from experience.

Unlike digital narratives, hypertext fiction consists of words, characters, plots—all the constituents of the Great Novel—and seems to present itself as narrative fiction's next leap, threatening a genre that is more than two hundred years old and that, according to Coover's "End of Books" and John Barth's article in a 1967 issue of the *Atlantic*, "The Literature of Exhaustion," can seem just about out of gas. The fear that hypermedia is what media ecologists once called a "killer technology," the equivalent of the Model T that ultimately supplanted the horse, accounts partly for the note of hysteria in some critical responses to hypertext. According to this view, the inclusion of Michael Joyce's *afternoon* in the *Norton Anthology of Postmodern American Fiction* signaled not just the arrival of the barbarian at the gate, but the presence of the barbarian on the living room sofa with its feet propped up and a hand already snaking out for control of the remote.[19] With their slow or jerky video clips, animation, and often clumsy interfaces, however, digital narratives threaten to obtrude on no such exalted tradition. While increasing sophistication in digital production and gains in hardware capacity and speeds have improved the quality of digital narratives exponentially during the past decade, there is still little risk anyone's going to mistake the likes of *Midnight Stranger* for *Chinatown*—or even for *Debbie Does Dallas*.

Yet for all their apparent differences, both hypertext fiction and digital narratives function like any medium in what historian Elizabeth Eisenstein calls its incunabular stage—an evolving form that in its infancy absorbs the media and genres that preceded it.[20] Both forms contain recognizable genres still borrowed from print, with digital narratives focusing primarily on popular genres: adventures, fantasy, mystery, and science fiction. Similarly, hypertext fiction mostly follows the path of late-twentieth-century fiction, characterized by multiple perspectives and voices, episodes linked with associative logic and memory, and rejection of the conventional, often pat, final awarding of marriages, happiness, money, and recognition that wrap up narratives in mainstream and genre fiction alike. Both forms speak, potentially, to questions that remain valuable, regardless of whether we pose them of works in print, on a multiplex screen, or on a flickering fourteen-inch monitor. What holds a narrative together, once you subtract its successive paragraphs and sequential pages or its linear scenes and sequences? What sorts of stories lend themselves to a medium in which readers can return to the same narrative—as they can with the likes of *Myst*—for more than forty-five hours without exhausting its full range of possible developments and outcomes? How will interactivity change the stories we tell in future? If a story has no physical ending, how do readers know when it—or, at any rate, they, are finished? What makes a narrative enjoyable? And, finally, why we read for pleasure—a question that critic Wolfgang Iser acknowledges in *The Reader in the Text* "has so far been barely touched on. If it is true that something happens to us by way of the literary text and that we cannot do without our fictions—regardless of what we consider them to be—the question arises as to the actual function of literature in the overall make-up of man."[21]

The chapters that follow aim to tackle some of these questions by forging comparisons between avant-garde print and hypertext fiction, examining readers' encounters with hypertext and with chunked print narratives to inspect the invisible, intangible "glue" that can cement even discrete pieces of text together, and scrutinizing, in particular, what happens to our sense of stories when they lack the closure of print narratives. While *afternoon, WOE,* and *Victory Garden,* the fodder for some of these explorations, can no longer be considered isolated examples of substantial hyperfictions, they remain some of the most complex in the medium, due, in part, to the number and types of links featured in all three works, as well as to conditional links and cognitive maps not yet available in hypertext fiction on the World Wide Web.[22] The version in chapter 4 of Stuart Moulthrop's "Forking Paths" that so tormented my class at New York University in the mid-1980s represents an isolated case of a roomful of canny readers who'd never before so much as *heard* the word *hypertext* and an author's intentions in creating a structure for interaction that followed what seemed like sound theories about how readers interpret print literary texts—but sheds significant light on the accuracy of theories of reading when readers discover how some theories of reading print texts may have entirely different effects and consequences for readers navigating through hyperspace.

The pages to come also focus on the aesthetic, cognitive, and physical aspects of reading actual interactive narratives—partly to address the current dearth of such studies, but, more important, to explore what happens to readers' comprehension of and pleasure in fiction when narratives have no singular, physical ending. By examining in detail both the similarities and differences between interactive and print stories, we can begin to understand the satisfactions we derive from being drawn into fictional worlds not of our inventing— one of the opportunities afforded us when we encounter reading, stories, plots, and characters outside a print environment so familiar to us that we are scarcely aware of print as both medium and technology. By bringing together disparate studies in the fields of psychology, narratology, artificial intelligence, and literary theory, we can begin to understand which elements of storytelling are changeable, open to further development and invention in interactive narratives, and which are changeless and immutable across media and millennia alike. We might even, perhaps, anticipate the look and feel of narratives in the future when we discover for ourselves the adrenaline rush of navigating between Scylla and Charybdis as we wend our way through a twenty-first century simulacrum of the tales of brave Odysseus.

Books without Pages—Novels without Endings

Undoubtedly man will learn to make synthetic rubber more cheaply, undoubtedly his aircraft will fly faster, undoubtedly he will find more specific poisons to destroy his internal parasites without ruining his digestion, but what can he do mechanically to improve a book?
—Vannevar Bush, "Mechanization and the Record" (1939)

What if you had a book that changed every time you read it?
—Michael Joyce (1991)

During the spring of 1998, Flamingo Press began selling softcover versions of the World Wide Web hypertext fiction *253*, Simon and Schuster released two versions of *Douglas Adams' Starship Titanic*, one, a novel in hardcover, written by former Monty Python member Terry Jones from a story by Douglas Adams, the other, an interactive narrative based on the same story that shipped on three CD-ROMs. As the reviews trickled in, the interactive narrative was roundly applauded for its intricate design, lavish graphics, and a cast of characters comprised nearly entirely of animated "bots" (or robots), programmed to respond to words and question marks in the users' input with over ten thousand responses—or roughly sixteen hours of spoken tidbits.[1] The novel, conversely, was roundly thrashed by most reviewers, who characterized it as "pretty thin and familiar fare."[2]

At the same time, as public interest in the fate of the *Titanic,* her passengers, and crew rose with the release of the blockbuster film *Titanic,* visitors strolling through the exhibition of the same name at the Florida International Museum in St. Petersburg crowded around a monitor in a gallery that featured one of the wrought iron windows from the ship's Verandah Café, china used on the maiden voyage, and corked bottles of champagne recovered by salvage teams after nearly three-quarters of a century on the ocean floor. The images on the monitor, however, drew more viewers than the artifacts themselves, displaying a rapid fly-through of the ship's decks and public rooms, all depicted in meticulous detail faithful to the original. Visitors who purchased the CD-ROM they had just seen in the museum gift shop, however, discovered that the fly-through of the ship was merely a small part of the digital narrative *Titanic: Adventure out of Time*. Published in 1996, *Titanic* features a cast of characters that are alternately drawn directly from the liner's passenger list, sly parodies of Gilded Age robber barons, or historically accurate approximations of the socialites and steerage passengers who might have boarded the White Star liner. As *Titanic: Adventure out of Time* websites encouraged users to tally the number of historical inaccuracies they encountered, newspapers released findings by a marine forensic team and metallurgist pointing out the unusually high concentration of slag in the steel used for *Titanic*'s rivets could have caused them to give way after collision with an iceberg, resulting in steel plates in the ship's hull springing open to the sea, a discovery confirmed during dives that revealed six thin holes in the bow, commensurate with opened hull seams.[3] Prominent among the subplots in the interactive narrative is an intrigue involving a blackmail scheme by an Irish serving girl in steerage who possesses a letter written to an Andrew Carnegie surrogate named Andrew Conkling, a letter that warns of the inferior steel manufactured at one of Conkling's mills and subsequently used in the construction of the *Titanic*'s hull—a plot woven into the narrative at least two years prior to the newspaper revelations about the rivets and the modest size of the holes in *Titanic*'s bow.

Readers enter the narrative via a flashback that begins with a close-up of a model of the *Titanic* in an otherwise dilapidated London apartment, seen during World War II's London Blitz.[4] Here you can rummage through the flat's scant contents, serenaded by the Cockney landlady's threats of eviction heard through the flat's closed front door, and discover a series of postcards that seem to document the central character's downward spiral toward dissipation. Or you can inspect the contents of the kitchen cupboards, complete with scurrying mice, look out the window at the barrage balloons tethered above London— and wait for the flat to take a direct hit from a Luftwaffe buzz bomb. The ensuing fire, in the mysterious means employed by adventure stories in both books and films, situates the user back in a stateroom aboard the *Titanic* on April 14, 1912. It is 9:30 P.M.: you have slightly more than four hours to wend your way through a series of tortuous plots and subplots, deciding which to follow and which to bypass, before the ship begins her plunge to the ocean floor.

Eight possible conclusions finish the narrative—and it turns out that, among the artifacts you need to barter for, recover, steal, or kill for, following the usual CD-ROM adventure conceit, are a painting by Hitler, a rare copy of Omar Khayyám's *Rubáiyat*, a notebook containing the names of Lenin, Trotsky, and other Bolshevik revolutionaries, and a diamond necklace. If you manage to hit on one of the strategies for getting off the ship in time with all the objects, you succeed in changing the course of history. As a British Secret Service agent, you presumably turn the notebook over to your superiors, the Bolsheviks are assassinated, and the Russian Revolution never occurs. Serbian revolutionaries are deprived of the necklace and rare book with which they planned to finance the assassination of Archduke Franz Ferdinand, thus halting one of the principal triggers of World War I. And, as the only painting to be recovered from the now-sunken liner, the work you bring with you in the lifeboat vaults an obscure, talentless painter named Adolf Hitler to public acclaim, neutralizing one of the forces that contributed to World War II. Seven other outcomes, each with alternative histories, are also possible, with the reader discovering the implications of his or her performance during a flash-forward that follows the sinking of the *Titanic*. With its elegiac music, eight decks of public rooms and staterooms for nosing around at your leisure, and well-written characters who do not necessarily assist your progress toward the endgames—a welcome change from the narrowly purposive characters and trajectory of most CD-ROMs, which can feel like old video arcade games with a little more context tossed in—*Titanic* can feel as immersive as an absorbing film or novel, engaging readers in the pleasurable, trancelike state Victor Nell in *Lost in a Book: The Psychology of Reading for Pleasure* defines as "ludic," the sense of becoming so immersed in a narrative that we become "lost" in it.[5] In the Age of Obsolescence, in which the average book or film can be consumed in a matter of hours, run-times for interactive narratives like *Titanic* and *The Last Express* can last as long as a regular working week, leading at least one critic to plead for "an 'adult' mode to games for those with limited time and patience."[6]

One of the purposes driving interactive narratives, however, is the desire for the inexhaustible story, the mystery that unspools with a fresh cast of suspects instead of gliding quickly through its denouement to a limited conclusion; the endless fount of stories that spring from Scheherazade during her 1,001 nights; the seemingly limitless versions of long-familiar tales that Homeric rhapsodes spun in response to cues and demands from their audiences. And it is precisely this vision of the book that never read the same way twice that

Michael Joyce claims prodded him into collaborating with Jay Bolter on the hypertext application that became Storyspace, as well as into work on his hypertext novel *afternoon*.[7] For all that, the book that changes every time you read it, responding to your moods, your whims, your latest fetish is, perhaps tellingly, a fantasy that has never been explored in print—unless, of course, you count the nightmarish, endless book in Borges's short story "The Book of Sand," which so curses its owner with weeks of sleeplessness as he tries to chart its limits that he finally slips it into the bowels of the Argentine National Library—presumably the same collection of tomes over which Borges himself presided during his term as the library's director.[8]

There is, of course, a perfectly sound reason why the kaleidoscopic book has never been attempted: part of the concept of the book is bound up in its fixity, the changelessness of its text. But if it were possible to have stories that really did interact with your choices, they probably would not have the singular endings we are familiar with from countless novels, stories, films, and plays. Part of the pleasure of reading them would come from your ability to decide, say, to execute Charles Darnay and save Sydney Carton in *A Tale of Two Cities*—or to make Philip Carey stop acting like such an utter wimp in *Of Human Bondage,* which means that closure would be something *you* determined, not Dickens or Somerset Maugham. So how would you decide when you were finished reading? And how would you know you were finished with the story for good, especially since your future readings could potentially trigger new twists and fresh narrative possibilities?

Even if you became used to reading this way, it is hardly likely that digital media like hypertext are going to supersede books, regardless of how much critics like Miller or Birkerts fret over the fate of the book and *le mot juste.* Radio and cinema went foraging for slightly different niches once television debuted on the scene, and ballooning numbers of video rentals, airings on premium cable and satellite channels, and pay-per-view showings have all helped recoup losses for films that were absolute dogs at the box office—an unexpected boon for Hollywood. It is hard to imagine books becoming the horse of the twenty-first century—a possession that has lost so much of its utility that only the well-to-do can afford to have one around anymore.

This is especially true since the book as a technology evolved over the course of hundreds of years through innovations like spacing between words, tables of contents and indices, standardized spelling and grammar, the development of genres and conventions, and, ultimately, copyright, aimed at shoring up author's rights and royalties, but which also ensured that readers encountered the author-ized ver-

sion of a work and not a hastily pirated copy that more or less repli-
cated what some quasi literate had made of the author's work.[9] Hypertext, conversely, has been with us only since the late 1950s in proto-
type and only for the past fifteen years in mostly primitive
applications that offer readers the sketchiest of notions of the contents
of nodes, the destinations of links, of where they have navigated
within a network of nodes and links, even of how much of the narrative they have consumed.[10]

If the book is a highly refined example of a primitive technology,
hypertext is a primitive example of a highly refined technology, a technology still at the icebox stage. This is a far cry from the zero-frost
refrigerator-freezer version of the technology envisioned by Aarseth
and Murray, who anticipate, respectively, machine-made stories and
frame-based authoring systems enabling writers to cycle through "the
possible plot possibilities, eliminating many of them and specifying
appropriate choices or priorities for situations where the story pulls
from multiple directions."[11] Further, while some genres may migrate
to the new technology and undergo sea changes as interactive narratives, others will remain best suited to book form. Countless directors
have taken cracks at turning the novels of Jane Austen into films, yet
there is little evidence to suggest that, say, the *Pride and Prejudice*
production that first aired in 1996 on the Arts and Entertainment network killed off sales of the novel; if anything, the television adaptation
bolstered book sales by introducing new audiences to Austen's acerbic
wit. And painstaking attempts by both Stanley Kubrick and Adrian
Lyne to adapt *Lolita* to celluloid only reinforce how much more pow-
erfully the voice of Humbert Humbert speaks to readers when it exists
as mere marks on a page than when it is brought to life by actors,
voice-overs, and vivid vignettes. In the entertainment industry, killer
technologies generally make only equipment obsolete: the compact
disk destroyed the market for turntables and vinyl alike but did not
alter so much as a single musical genre. The issue here is not whether
the book and interactive narrative can exist comfortably together, so
much as whether future readers will begin reading print works differently from the way they do now. If you had become accustomed to
seemingly inexhaustible books that altered according to your curiosity
or whims, how would you react once you returned to reading print
fiction like *The Postman Always Rings Twice*, or even *Ulysses*, where
the text just kept saying the same thing, no matter how many times
you turned to it?

For some, these scenarios and questions alike sound far-fetched,
the sort of futuristic musing that had us believing, back in the seven-

ties, that we would be rocketing from Washington, D.C., to New York on ultrafast trains by the early 1990s—when we are all too aware that passenger trains are still trundling along at speeds that would have been decidedly unimpressive during the Eisenhower administration. But these are legitimate questions one could ask of readers and writers of readily available works of hypertext fiction. Like print fiction, hypertext stories can enthrall, frustrate, amuse, repel, and even frighten their readers. Certainly, readers of print fiction like *The Good Soldier* may interpret what they read completely differently their second time through the novel, after having already learned that the narrator is quite possibly the world's most literal-minded reader of events and what he reports seldom a reliable indicator of what happened. If you were to read Michael Joyce's *afternoon, a story,* one of the first pieces of hypertext fiction ever written, you might also interpret things entirely differently the second time around. But what you were reading—the actual, tangible text itself—would also probably be different from what it was when you first opened the novel. It is not a matter of the river being different each time you cross it so much as it is a matter of your stepping into an entirely different river with each journey you take.

As Michael Heim has pointed out, the prefix *hyper-* can signal a hitherto undiscovered or extra dimension, so the term *hypertext* describes a tool that lets us use the printed word as the basis for a technology that considerably extends writing's reach and repertoire—mostly by removing text from the single dimension it has on the printed page.[12] Within the confines of hypertext, narratives consist of discrete segments of text, some of which may be read or experienced in what may seem like sequential order even when they appear in radically different settings. Hypertext fictions like *afternoon* feature multiple links for every segment of text woven together in a complex web of relationships, associations, and alternative constructions of what might have happened on the morning when the narrator, passing a roadside accident on his way to work, discovers he cannot locate the whereabouts of either his estranged wife or young son. Segments of text within a hypertext—referred to by critics as *pages, sites, windows,* or *places*—can be as brief as a word or image, or as long, really, as a short story. These are associated with other segments of text, not by their sequential order on a page, but by links, pathways through the text that can be created, in some circumstances, by both readers and authors. Readers following Matthew Miller's "Trip," a World Wide Web–based riff on the seventies road movie, can skip their way across the narrative by using a map of the continental United States, deciding which state to visit as they follow a hapless narrator who has ended up as caretaker to his ex-girlfriend's two kids and careens around the country in a fruitless effort to rendezvous with their transient mother, who is never quite where she is supposed to be. Within each state, readers can choose links designated by interstates, highways, and county roads, or follow arrow links indicating continuous events in a single scenario. No matter where you click on the United States, you will find the narrator, Jack, and Jill in scenes that retain their comprehensibility and integrity no matter what order they are encountered in. Follow the narrative through Florida and Texas or Michigan and Illinois, and you will always find the threesome, embroiled in some comic or vaguely threatening scenario, usually down to their last dime.

The organization of "Trip" recalls Roland Barthes's schema for *The Pleasure of the Text,* where he uses alphabetic order to place observations in discrete segments of text in what he felt was as close to a random order as possible.[13] Other interactive narratives, however, rely on causal sequences that branch into a variety of mutually exclusive outcomes. In some readings of "I Have Said Nothing," for instance, readers encounter a punitive mother who whisks away the artifacts her son keeps beneath his pillow—keepsakes from his dead girlfriend that include a segment of her skin—but in others, his mother protects him by following to the scene of his girlfriend's accident and preventing him from committing suicide there. This difference between interactive and print narratives can make comparing accounts of what each reader thought the story was "about" in a literature seminar infinitely more varied and problematic for interactive than for print texts, since the events in the story itself, as well as its interpretation, may differ between students' experiences of the text. While authors of print narratives can never be certain exactly how readers will interpret their fiction, authors of interactive narratives can occasionally be surprised at the permutations and combinations of narrative segments that readers encounter—especially since hypertexts with more than a hundred segments and two to three hundred links will generate hundreds of possible versions of the text, some of which the authors themselves have neither anticipated nor seen.

Books without Covers

It is not that the Author may not "come back" in the Text, in his text, but he then does so as a "guest." . . . He becomes, as

it were, a paper-author: his life is no longer the origin of his fictions but a fiction contributing to his work.
—Roland Barthes, "From Work to Text" (1971)

The reader is the space on which all the quotations that make up a writing are inscribed without any of them being lost; a text's unity lies not in its origin but in its destination.
—Roland Barthes, "The Death of the Author" (1968)

Of course, critics, many of them steeped in literary theory, have begun noticing affinities between the features of hypertext and the way that poststructuralist theorists had described the Text.[14] Even the neologisms coined to describe print text—*liaison* (link), *toile* (web), *réseau* (network), and *s'y tissent* (interwoven)—seem, uncannily, to anticipate the hallmarks of hypertext. You do not need to be particularly perceptive to see the possibility of producing sequential yet nonlinear discourse with hypertext as an illustration of Jacques Derrida's contrast between linear and nonlinear writing. Nor do you need to be fantastically well versed in the writings of Roland Barthes to recognize hypertext in his description in "From Work to Text" of print text as a network of references to and reflections of other works. Or to seize on Derrida's definition of text as "a differential network, a fabric of traces . . . [overrunning] all the limits assigned to it so far"[15] as a decent sketch of a hypertext like Christiane Paul's *Unreal City: A Hypertext Guide to T.S. Eliot's* The Waste Land, where readers leap seamlessly between extracts of Jessie L. Weston and snatches of Chaucer, from Ezra Pound's contributions to its editing into the familiar, canonized version of the poem, even earlier draft versions of the work, as they traipse through lines of T. S. Eliot's *The Waste Land*—without needing to search through catalogs, wander library stacks, or page through dozens of sources. If activities like these felt familiar the very first time you surfed the World Wide Web, it may well be because you have encountered something like this already in the works of Barthes, Derrida, Foucault, and J. Hillis Miller, theorists who see meaning as distributed within texts through dense networks of associations with other texts reaching past the boundaries of the physical work itself. Or you might be reminded of the relatively recent shift in critical attention from authors and texts to the role of the reader, prefaced by Barthes's famous "Death of the Author" that declares the reader is the single device that ultimately controls the signifying potential of any text.

When you read your first hypertext, your first reaction might be that the Author has not only been killed outright, but that he or she has been quicklimed. It can be difficult to detect anything that even *smells* like an author or so much as a consistent editorial viewpoint amid the tangled network of texts and links. In, for example, *A Reader's Guide to "The Waste Land,"* lengthy discussions of the influence on Eliot's poem of *From Ritual to Romance* and *The Golden Bough* sit alongside extracts from the works themselves, as well as Pound's dismissal of both the footnotes and Eliot's conviction that the notes supplied readers with information crucial to a full appreciation of the work. Here, also, in Paul's hypertext—as in George P. Landow's *Dickens Web,* and Landow and John Lanestedt's *The "In Memoriam" Web*—critical essays by students and contributions by present-day professors of literature share the same continuous space with a plethora of other voices, including, in *The Dickens Web,* the words of luminaries of Victorian culture. This should hardly astonish anyone who has already encountered the concept of intertextuality or the decentered text. After all, one of the hallmarks of Barthes's notion of text was its plurality, its polyvocality, the legion of voices with which it spoke. But whereas this chorus of voices is strictly implicit in the printed realm, a kind of dragon that every writer faces down (with varying amounts of sweating and dexterity), it is virtually omnipresent in hypertext. By isolating segments of an argument we would normally digest in physically discrete places with a single, unthinking gulp, writers in hypertext can open up apertures into conflicting and even mutually exclusive perspectives.

The polyvocality of hypertext does not mean that it either imitates or induces a kind of schizophrenia in its readers. Since hypertexts are fluid where print texts are fixed, the medium lends itself to circumstances where readers can play out alternative scenarios, even experience all possible outcomes stemming from a single set of circumstances, as in, for example, Shannon Gilligan's Multimedia Murder mysteries, particularly *The Magic Death*—an interactive narrative that invites you to play detective and nail multiple murderers who have vastly different motives (and methods) for knocking off the same victim, who, it turns out, can appear to have been bumped off by various killers in three distinct but equally feasible scenarios. When it comes to building theories of, say, how an accident happened, or how society and technology come together to produce the tools we use, the printed word—static, linear, and relatively austere—does not necessarily stand out as the best way of doing justice to the complexity of experience with all its contingencies and possibilities. Originally fashioned for use in the law courts of ancient Athens as a tool of persuasion, the rhetorical tradition that has most strongly shaped Western print conventions turned writing into a way of reducing multifaceted and indeterminate experience into singular and linear representations of events that removes from them as much ambiguity as possible.[16] Print mostly works in much the same way as a legal decision: a zero-sum game that settles the conflicting claims and elaborate narratives constructed by each side with a single decision, inevitably validating one version of events entirely while suppressing the other. Today our representations of the world have been shaped by philosophical relativism and, in areas like quantum physics and chaos theory, we increasingly embrace a world infinitely more complex, unpredictable, and indeterminate than anything our nineteenth-century forebears could imagine. Yet we still overwhelmingly rely on a medium that, when we bow to the dictates of its conventions and rhetoric, makes us all, more or less, into objectivists or positivists, regardless of our intentions.

These issues can partially account for hypertext's having become a cultural buzzword of sorts—a noun everyone from hard-nosed journalists to sociologists and even professors of English appears on nodding terms with. Media theorists like Marshall McLuhan and Walter Ong have dedicated entire careers to tracing the roots of cultural developments as seemingly disparate as the concepts of intellectual property and universal suffrage to their origins in the moment when Gutenberg adapted an old winepress to the printing of Bibles. So we should not be startled that the arrival of a technology apparently promising a radical shift in the relationship between text, reader, and world—one that has been proclaimed as "the first *literary* electronic form"[17]—should be heralded with so much critical attention that critics complain about the disproportionate relationship between the preponderance of articles *on* hypertext fiction and the slender number of actual examples floating around.

These larger claims, nevertheless, can help shed light on the alternatives to print represented by hypertext—and on its potential impact on the domain of both readers and writers. Not a few critics writing on hypertext fiction have already pounced on the similarity between the historical moment described in Plato's *Phaedrus,* where Socrates disparages writing as a form of representation, and our own, where the arrival of hypertext enables us to deride the printed word for roughly the same reasons:[18]

> The fact is, Phaedrus, that writing involves a similar disadvantage to painting. The productions of a painting look like

living beings, but if you ask them a question they maintain a solemn silence. The same holds true of written words; you might suppose they understand what they are saying, but if you ask them what they mean by anything, they simply return the same answer over and over again. Besides, once a thing is committed to writing, it circulates equally among those who understand the subject and those who have no business with it; a writing cannot distinguish between suitable and unsuitable readers. And if it is ill-treated or unfairly abused it always needs its parents to come to its rescue; it is quite incapable of defending or helping itself.[19]

To someone steeped in a primarily oral tradition, the written or printed word has obvious drawbacks: it is singular; it is static; and no matter how many times you read something (or how your purpose shifts from reading to reading), it just keeps saying the same old thing. Between Plato's era and our own, of course, poststructuralist criticism has made these sentiments seem a trifle quaint, and, if the physical text itself is unchanging, our readings of the same text are far from it. My reading of *Flaubert's Parrot* will seem different if I am after a good read than if I am pursuing snippets of critical or biographical wisdom on Flaubert, just as none of us will read Louis Althusser's *Pour Marx* in the same way again after we stagger through *The Future Lasts Forever* and discover there that, as far as Althusser was concerned, a little of Marx went a long way. Rereading Plato is, in fact, no different from either of these cases. Although the text is the same, the sentiments expressed in *Phaedrus* jump from seeming merely historically interesting to strikingly current when I read them after first discovering hypertext.

Books without Endings

Writing, when properly managed . . . is but a different name for conversation. As no one, who knows what he is about in good company, would venture to talk all. The truest respect which you can pay to a reader's understanding, is to halve this matter amicably, and leave him something to imagine, in his turn, as well as yourself.
—Laurence Sterne, *The Life and Opinions of Tristram Shandy* (1759)

Nothing odd will do long. *Tristram Shandy* did not last.
                                    —Samuel Johnson (1776)

Plato, of course, did not remain alone in his mixed feelings about writing (although contemptuous of it, he did, nonetheless, record his dialogues with it) until Vannevar Bush began tinkering around with his plans for the Memex in the thirties. Since print is linear, fiction writers generally also relied on straightforward, chronological narratives for telling their stories from the infancy of print fiction on through the tail end of the Victorian era. Yet if we turn to Laurence Sterne's *Life and Opinions of Tristram Shandy,* we can watch a novelist continuously (and, to his contemporaries, doubtlessly outrageously) playing with, poking fun at, and occasionally doing violence to the constraints of the medium—activities that seem to have more in common with hypertext fiction than they do with the literary conventions of Sterne's day. Ostensibly, Sterne's novel is a bildungsroman—only this first-person narrative attempts to begin with Tristram's conception, not his actual birth. The plot almost immediately grinds to halt as Tristram sets the exchange between his parents in the context of his father's relations with the Shandy family and the prenuptial agreement between his parents that leads to Tristram's mother giving birth down in the country that will lead to the regrettable flattening of Tristram's nose during his delivery . . . The narration of events in a seamlessly linear, chronological fashion is, for Tristram, well-nigh impossible: his retelling of events leaps forward associatively, triggered by similarities in words as well as by memories—a foretaste of the associative power of Proust's madeleine. Each time Tristram attempts to smooth out the jumble of causal relationships and associative episodes into the straightforward, linear story his readers expect, the interpenetration of circumstances, the workings of the mind, and the limitations of the printed text all conspire to snarl them up again.

Was Sterne merely playing with conventions and readers' expectations—or was he already chafing a bit at the confines of the printed page and trying to joke his way past them? Mimicking the epistolary novels of his era that were directed toward a reading subject (who was not, however, the "real" reader but a character within the fiction itself), Tristram directly addresses his surrogate reading public, but in the form of a ridiculous construction—a slightly stuffy female reader. Taking her assumptions at face value, Tristram uses his projections of her responses as jumping-off points in his storytelling, just as he later includes a chapter on whiskers because he claims he had promised it to his reader, as well as a blank page for her to sketch in the features of

Widow Wadman. As Sterne's novel progresses, the narrative occasionally seems to push back the limits of the physical text. Chapter 24 is missing from book 4 because, as Tristram explains in chapter 25, he has torn it out of the book. He then proceeds to recap everything we "missed" in the lost chapter, yet our sense of that "loss" is hard to entirely dispel, since the numbering of pages in chapter 25 continually reminds us that we have skipped a chapter. Elsewhere, Sterne teasingly follows up the misunderstanding between Widow Wadman and Uncle Toby over his war wound—Uncle Toby's promise to show her "something" in the garret has her, not surprisingly, anticipating something else entirely—with two blank chapters, leaving the reader more than a little something to imagine.

Where we might have read a novel like Richardson's epistolary *Pamela* quite comfortably from our position as a voyeur-reader, Sterne's novel makes the whole convention of the fictionalized reader seem faintly ridiculous and, at times, downright inadequate. In constructing a narrative to mimic the conversation he knows the print novel cannot hope to emulate, Sterne reminds us of just how static and fixed print fiction can be. Compared with the interactions required by conversing or reading hypertext fiction, reading print can seem a tad like listening to a monologue or a lecture, where you basically have two choices: listen or leave. While reading hypertext fiction hardly involves the same order of dynamic interaction encountered in conversation (for more on this, see chapter 3), you can interrupt the flow of the narrative or exposition in search of more congenial pastures, leap between conflicting representations of a single subject, dwell a bit more intensively on a moment or topic—and skip others entirely. A work of hypertext fiction can act as a blueprint for a series of potential interactions, and your movements through it, a dance choreographed by an absent author who has anticipated the questions, needs, and whims of imaginary readers. It is an encounter and a performance not terribly distant from the kind shared by ancient rhapsodes, who sifted through memorized, formulaic versions of cherished stories, and the live audiences whose demands partially shaped the versions they heard.

So where, exactly, does this leave the reader: behaving something like an actor improvising within an assigned role in a John Cassavetes film—or like a reader enjoying a mild frisson of pleasure at being able to choose an ending in one of those "Choose Your Own Adventure" stories? Since hypertext fiction does not have the fixed, tangible beginnings and endings of print stories and books, readers decide where their experience of the text ends. While examples exist of print stories or novels with multiple endings—think of stories like "The Babysitter" in Robert Coover's *Pricksongs and Descants* or John Fowles's *French Lieutenant's Woman*—there is only one episode of closure that can actually end the thing. Whatever comes last tends to seem like the "real" ending, and what reader would close the book after the "ending" encountered midway through Fowles's novel with a bulging stack of pages and chapters acting as a reminder of what he or she is missing? A hypertext story, on the other hand, can feature as many points of closure as its author can dream up: one, eighteen, or eighty. Which of these points becomes the ending to the story, the place where a reader feels he or she can fold up the tent, so to speak, and be done with it, is entirely up to the reader.

Yet this is no simple "Choose Your Own Adventure" scenario where readers can see for themselves that they have exhausted whatever possibilities the narrative held and pass the book on to their friends. Since hypertexts can include hundreds or even thousands of narrative episodes or segments, connected with an even vaster number of links—and each of these bridges between texts can require readers to satisfy specific conditions to traverse them—a single work of hypertext fiction can have thousands of permutations. As a reader, you can return to the same narrative over and over again, never entirely certain what will happen with each new version of the text you realize. As an author of a hypertext fiction with several links for each segment of text, you yourself may never experience all of its possible permutations, providing you with only a modicum of the control you possess on the printed page over the ways in which your text will be actualized once read. The physical text can change with each reading and reader—so texts can behave unpredictably even during events normally as tame as a public reading. Picture a reading session of the usually unremarkable kind that prefaces book signings where, however, the hypertext writer has roughly the same odds on predicting what she is going to read next as you might enjoy on a little wager at the track, and you can begin to imagine just how radically hypertext fiction can reconfigure the roles of reader and writer alike.

Hypertext—Where Technology and Theory Meet

Hypertext is as much a concept as it is a form of technology. As a technology, hypertext shot into public awareness in the late eighties when Apple began shipping copies of its own hypertext software, Hyper-

Card, with every new Macintosh—although comparatively few HyperCard users were aware they could write hypertext stories or documents with it, since HyperCard was also an all-singing, all-dancing utilities package that boasted clip art, a drawing and painting program, and the digital equivalent of a Rolodex. Individual units known as "cards," containing text or images or even sounds, could also be organized into stacks and the cards linked together in a variety of sequences. But the component that distinguishes hypertext from your average database—the capacity to link segments—did not reveal itself easily to less than adroit HyperCard users, since cards could be linked only when writers used HyperTalk, HyperCard's own built-in scripting language, to forge them. Other hypertext software, most notably Guide and Storyspace, was to prove more congenial for would-be hypertext writers—particularly Storyspace, which enables writers to create sophisticated links and attach guardfield conditions to them by simply pointing and clicking. By the late eighties, the first hypertext stories began circulating mostly on a samizdat basis between users, with a speed and efficiency that sometimes startled their authors. While still writing *afternoon* years before its publication, Michael Joyce found himself bombarded with questions from a reader who had managed to lay her hands on a copy of the novel—try to envision that happening with Hemingway working on an early draft of *The Sun Also Rises,* or any other print author you care to imagine.[20]

However recent a technological development hypertext may be, as a concept, hypertext has been rolling around for decades. The idea was born in the thirties, the child of a science adviser to Franklin Delano Roosevelt, Vannevar Bush, who envisioned a system that could support and improve human memory more efficiently than the printed word. Scientific research, Bush claimed, was being bogged down by the mechanisms used for storing and retrieving information,

> largely caused by the artificiality of systems of indexing. When data of any sort are placed in storage, they are filed alphabetically or numerically, and information is found (when it is) by tracing it down from subclass to subclass. . . .
>
> The human mind does not work in that way. It operates by association. With one item in its grasp, it snaps instantly to the next that is suggested by the association of thoughts, in accordance with some intricate web of trails carried by the cells of the brain.[21]

If Bush felt that books and libraries had begun to hinder research as much as they helped it, part of the impetus for his building the Memex also stemmed from a belief that machines could model the processing of information by reproducing the neural structures in the brain that linked information together by association rather than by the linear logic of the printed book.[22] Readers using Bush's Memex would collect snippets of information from a huge variety of sources, link them together with "trails," and even insert comments or notes of their own. New kinds of reference materials using the same technology as the Memex could be seamlessly incorporated into the wealth of available information, to which readers could add their own, personalized Memex trails.[23]

Ironically, the most striking aspect of the Memex—its potential to radically reorganize the valuing and handling of information as a commodity—was treated by Bush as simply part of a system that made storing and retrieving information more efficient. Nothing in Bush's writings on the Memex remotely addresses the implications of readers customizing and distributing their own versions of the new encyclopedias and reference works he envisions—or the knotty problems that could arise if, say, a user had circulated a highly idiosyncratic and heavily annotated version of *The Oxford English Dictionary* that became the de rigeur version of the OED for thousands of readers. If we assess his probable response based on the strong streak of technological utopianism in Bush's other writings (particularly "The Inscrutable Thirties"), however, we could guess that Bush would have been taken aback by the very question, and, particularly, by the notion of information as a commodity. His invention was, after all, intended to facilitate the sharing and propagation of information to promote scientific progress—perhaps not coincidentally, the same impetus behind the founding of the World Wide Web in the early nineties. Further, Bush's writings and their insistence on making the production of research more efficient may also have taken their cues from a more pragmatic source: the drive toward progress begun during the Second World War and continued during the Cold War era, when scientific research was spurred on primarily by America's adversarial relationship with Germany and, later, the Soviet Union.

Revolutionary potential aside, however, the analog machines that Bush had slated to form the basis of the system were soon superseded by digital technology, and, as a result, the Memex was never built—although Bush's articles served as a direct inspiration for the work of Douglas Engelbart and others who proposed and eventually constructed digital approximations of the Memex. Not surprisingly, the unexpected by-product of Bush's vision—the concept of a medium that blurred the line between author and reader and between the contents of one work and another—was also to prove both potent and remarkably hardy. In the decades following the publication of Bush's Memex articles, the work of pioneers like Engelbart and Ted Nelson (Nelson dreamed up the label *hypertext* and became its relentless champion) ensured that the concept of a new technology capable of enhancing the roles of reader and writer alike hung on, frequently precariously, to the margins of the public agenda.

The Reader Comes of Age

The reader does not merely passively accept or receive a given literary work but through the act of reading participates along with the author in the creation of the fictional world evoked by the heretofore lifeless text. . . .

At first glance, interactive fiction acts out this process literally. It seems to emancipate the reader from domination by the text putting her in at least partial control of the sequence of events. . . .

. . . interactive fiction looks as though it acts out one particular model of reader response. Iser has suggested that the text of a novel lays down certain limits, but within those limits are gaps which a reader feels impelled to fill. An interactive fiction seems to make this arrangement explicit.

—Anthony Niesz and Norman Holland,
"Interactive Fiction" (1984)

Around the same time that Bush was reflecting on the inadequacies of printed books, indices, and card catalogs, literary critics were beginning to show the first signs of restiveness over the field's intensive preoccupation with the Text. If, in the history of literary criticism, the eighteenth and nineteenth centuries were the era of the Author, then the early twentieth century, with the advent of the New Criticism, belonged to the Text. The first stirrings of critical interest in the reader's share of the author-text-reader triumvirate dates roughly from the thirties, beginning with Louise Rosenblatt and Jean-Paul Sartre, and eventually spread rapidly through everything from structuralism to semiotics by the seventies. One of the reasons the era of the reader took so long to arrive at the forefront of literary criticism most likely lies in the notorious difficulty facing anyone who sets out to describe the act of reading. Texts are, after all, physical objects that can be scrutinized and cited. Authors produce tangible products and tend to be written about by historians or contemporaries—and to helpfully give interviews or even write extensively about themselves and their titanic struggles with this or that work. But the act of reading is invisible, aside from eye movements and the odd subvocalization here and there. We can assess the marks on the page and the white space around them. We can study what readers say or write afterward in response to prompts, probing questions, or even videotaped images of themselves mulling over the text—and still we cannot isolate what happens when a reader reads. It is a process that is part perception, part convention, and entirely stubbornly, unchangeably intangible. Although we can weigh, measure, and describe texts the way we do brains and neural functions, we can only guess about the way reading and interpretation functions, as we do minds and thoughts.

There are other difficulties, of course, facing the would-be critic attempting to nail down a definition of what, exactly, reading consists of. We can hedge a guess that it involves a certain amount of interaction with a text, which, as most theories of reading agree, works something like a blueprint for potential aesthetic experiences. In the theory of reading articulated by Wolfgang Iser in *The Act of Reading*, readers realize or concretize texts from skeletal structures on the page, since the written word—unlike the spoken—can never completely be tied to a single, determinate meaning. This is where things begin to become wretchedly complicated, making us inclined, perhaps, to recall Plato's complaints in *Phaedrus* more sympathetically. For example, I would be able to correct you if you misconstrued what I meant when I said aloud to you, "I get a nosebleed every time I go above Fourteenth Street." I could remind you that I prefer to spend most of my time in Greenwich Village, tell you that the northernmost boundary of the Village is Fourteenth Street—and point out that the nosebleed is strictly a figure of speech. On the other hand, if I were to include this sentence in an essay or piece of fiction, you might not understand just what I meant by the allusion to Fourteenth Street. Even if you assumed that I was referring to the Fourteenth Street in New York and knew that "above" meant "north," so that "above Fourteenth Street" meant leaving the Village, you might wonder whether I intended to say that the Village was a more Bohemian, or down-to-earth, place than the rest of the city, or whether I was saying that I found neighborhoods like Murray Hill or the Upper West Side a bit headier and more exciting. But you might not even understand that I was referring to the Fourteenth Street in Manhattan and not the Fourteenth Street of Brooklyn or Queens, also in New York. You could mistake my reference to a specific street in Manhattan and suppose that I was jokingly connecting the size of street numbers and their heights above sea level. Or you might even end up wondering just what it is in the air in Chelsea or around Gramercy Park that provokes nosebleeds.

Indeterminacy can make even a brief sentence into something over which we could easily spill a few paragraphs worth of ink in the interpreting. It is also the one feature that all theorists studying the act of reading agree is the most prominent characteristic of written texts: holes. For Sartre, fiction is rife with "traps" that lure the reader into an act of "directed creation."[24] For Roman Ingarden, one of the earliest of the phenomenological theorists, texts were paradoxically filled with "gaps" or "places of indeterminacy."[25] For Wolfgang Iser, who was influenced strongly by Ingarden, readers were provoked into fleshing out texts by "gaps," "blanks," "vacancies," and "negations."[26] Even theorists examining the process of reading from disciplines outside literary criticism—such as cognitive psychologist Roger Schank and psycholinguist Frank Smith—have claimed that reading is driven by readers' needs to fill in gaps or spots of indeterminacy in the text. This is, of course, part of what makes describing the act of reading so thoroughly baffling a task. Readers not only perform an invisible act; they also would appear to perform it largely by filling in holes in the text that generally do not even appear visible to most of us. We are so accustomed to making assumptions, filling in blanks, and inferring causes and effects as we read, in most cases, that they have become all but automatic.

This ease with which we perform this act daily (on everything from signs in the subway to the likes of *The Bridges of Madison County* or *The House of Mirth*), which helps to render reading well-nigh invisible, can account for some of the neglect the reader has suffered throughout the history of literary criticism. Conversely, it was the critics' recognition of just how little explicit information novels and stories supplied their readers that led to interest in theories of reading. If reading represented such a seething hive of activity, their thinking went, wasn't it rather creative? Sartre was among the first to think so:

Raskolnikov's waiting is *my* waiting which I lend him. Without this impatience of the reader he would remain only a collection of signs. His hatred of the police magistrate who questions him is my hatred which has been solicited and wheedled out of me by signs, and the police magistrate himself would not exist without the hatred I have for him via Raskolnikov. That is what animates him, it is his very flesh. . . . [T]he words are there like traps to arouse our feelings and to reflect them towards us. . . . Thus for the reader, all is to do and all is already done.[27]

"Reading," Sartre declared, "is directed creation," which sounds like it accords the reader a measure of freedom hitherto unacknowledged in literary criticism. But what he seems to be granting them is something more oxymoronic: a strictly limited freedom. He is, after all, talking about *directed* creation, which brings to mind a child let loose with a coloring book, provided with already finished pictures drawn out with thick, black lines and nice, inviting white space on either side of them. Dostoyevsky supplies Raskolnikov's hatred, and we bring along our memories of, perhaps, certain junior high school administrators we loved to loathe, or our recollections of the sleazy cops in *Serpico*.

There is also something telling in the language Sartre uses to stake his claim for the inventiveness of the reader's act. The author's collection of signs wheedles and solicits, tricking our emotions out of us like an escort posing as a date. Reading, interpreting, seems to be something we do almost in spite of ourselves. In fact, our contemporary use of the word *read* as a synonym for *interpret* (as in "Yeah, that's how I read his character, too") is rather telling: to read *is* to interpret. Like a carnal streak we cannot quite repress, we respond to the text's machinations the way one might to a come-on, but, according to Sartre, we end up having not so much a moment of interaction as an interlude in which we merely recognize our own emotions reflected back to us. Although Sartre insists the text leaves it "all to do," his "all is already done" conclusion undermines his claim for the reader's relative sovereignty by insisting that the text is, essentially, complete when we get to it. We are simply drawn into moments in the text and then trapped, coaxed into coloring them in with the hues of our own memories. It all seems just a bit like karaoke—no matter how you may sing it or what marvelous vocal flourishes you pump into the song, the melody will always be unchanged, the song never yours.

In its later incarnations, reader-response criticism attempts to crown readers with laurels far more exalted than Sartre's "directed creation." The great phenomenologists of reading—Roman Ingarden, Hans Robert Jauss, and Wolfgang Iser—all tend to see reading as rising out of readers' interactions with a text that can seem less an explicit string of instructions than a sketchy blueprint, inviting something tantamount to artistic license in the interpreting of it. In particular,

Iser—arguably the most influential theorist of reading—seems to stake out generous territory for the reader's share:

> [G]uided by the signs of the text, the reader is induced to construct the imaginary object. It follows that the involvement of the reader is essential to the fulfillment of the text, for materially speaking this exists only as a potential reality—it requires a "subject" (i.e., a reader) for the potential to be actualized. The literary text, then, exists primarily as a means of communication, while the process of reading is basically a kind of dyadic interaction.[28]

In a dyadic interaction, both parties play more or less equal roles, just like people engaged in conversation—an activity widely considered the model for what interactivity should ideally look like. So Iser's version of reading seems a far cry from Sartre's mere "directed creation," and more like the sort of act that might describe what readers bring to hypertext narratives. In print fiction, readers actualize a text by mostly unconsciously fleshing out adjectives, adverbs, and nouns, by making assumptions and inferences, and framing hypotheses about what is happening and what is coming next. Readers of hypertext fiction, though, perform something a bit more like an act of concretization, by blazing along trails through the dense web of possible hypertextual links, activating conditions with effects that the author may not have even anticipated. In this respect, hypertext has more in common with dance than it has with novels like *Play It As It Lays* or *Ulysses*. Until a reader assembles it, performing it, the text exists only as a set of potential motions, a sequence of steps and maneuvers that become actualized only at the instant that the reader selects a segment of text or fulfills a condition for movement.

Yet, as we find in Iser's theory of reading, hypertext readers are restrained by determinate things in the text, for all their uncertainty about what is going to rear up next. While it is possible for me to construe what happens in the same passage in *afternoon* entirely differently in each of the contexts in which I encounter it—as we will see in chapter 4—certain features of that segment of text remain constant. Although I may be invited to fill in details of what characters are thinking or mull over what hanky-panky might be percolating behind the scenes to inspire their badinage, I cannot interpret the word "restaurant" in that passage to mean "boardwalk" and animate the scene with my own memories of Coney Island: what I already know about the characteristics of restaurants and what people do in them shapes the content of what I can intelligibly project onto the restaurant scene.

In fact, according to some theorists, my ability to comprehend most of what I read has something to do with a knowledge of restaurants, as well as schools, banks, and lines at cash registers—a vast array of strategies for behavior in the real world. When I venture into a restaurant, I do not wonder what I am supposed to do with the piece of laminated cardboard someone in uniform shoves at me. Nor would I ask the server wielding a pen and notepad if he or she could find someone to take a look at my sick cat. Both my perception and my behavior are more or less directed by my knowledge of what are called, variously, depending on whether you are a cognitive scientist or an art historian, scripts or schemata. Generally, we are not conscious of calling on a script when a situation approximates a familiar one, however distantly, yet, according to schema theory, we use our experience of them constantly at a variety of different levels. In our experience of the world, schemata serve to guide us through driving in heavy traffic, endorsing checks, and even performing the Heimlich maneuver on a dining partner over lunch. When we read, schemata incline us toward forming hypotheses that lead to our making some inferences and excluding others without our ever being aware of the existence of latent, alternative interpretations.

Consider the following passage:

Tony slowly got up from the mat, planning his escape. He hesitated a moment and thought. Things were not going well. What bothered him most was being held, especially since the charge against him had been weak. He considered his present situation. The lock that held him was strong but he thought he could break it. He knew, however, that his timing would have to be perfect. Tony was aware that it was because of his early roughness that he had been penalized so severely—much too severely from his point of view.[29]

If you think that this passage is about the thoughts of a convict meditating on his escape, you probably focused on the words "escape," "being held," "the charge against him," and "the lock"—all the components of imprisonment, familiar to most of us through stories or films set in prison. If, however, you have a slender knowledge of the rules and regulations of wrestling, you will most likely have read this as a description of a wrestler trying to break free of his opponent's hold. Here, too, you would focus on many of the same features that enable you to latch onto the convict scenario, yet you would also probably have seized immediately on the presence of the mat in the first sentence. As readers and viewers, most of us are less than familiar with the wrestling script and, not surprisingly, unlikely to latch on to its cues.

Generally, when we use the jailhouse script to flesh out the details, we simply screen out the stray items here and there that do not fit—like the mat—without being aware that an alternative schema and, with it, an entirely different reading, exist. But if schemata represent a substantial part of what guides and restricts our experience of what we perceive, they help us only while we are operating squarely within the confines of scenarios we can recognize. Let's say, for example, that we are watching a film like *Jacob's Ladder*, one that shuttles rapidly from genre to genre, suggesting by turns that it is a war flick, a thriller about secret military plots, a drama about madness, and a horror film on the order of *Rosemary's Baby*. How do we predict what is going to come next? Which genre constraints does the film—and its ending—need to satisfy? Scripts can help nudge us toward comprehension at a local level only if we can recognize individual elements as part of a larger script. If this sounds like an endless loop—like the hermeneutic circle—it is because the process actually is one. What makes it work, what enables us to understand what we see or read, is the relative fixity of the larger script, as well as the script most of us possess telling us that films and novels tend to follow the single, overarching schema we would call a genre, pretty consistently throughout. Works that violate genre constraints randomly or haphazardly like *Jacob's Ladder* tend to frustrate us, since we need consistency to either identify a new schema or to broaden or modify our concepts of already existing ones. When hypertext readers attempt to build a script, however, even if the story adheres faithfully to our expectations of the genre, the local scripts can seem a bit mercurial. Already, some hypertext narratives, such as *afternoon* and *WOE—Or What Will Be*, have capitalized on our tendency to project scripts and interpret everything we read in light of them (as we will see in chapters 4 and 5), so that you can be jolted when you discover the "she" and "he," whose identities you thought you had pegged correctly from the start, turn out to have shifted from place to place, referring in each segment to entirely different people.

The reader, we have seen, does not wait until the end to understand the text. Although texts provide information only gradually, they encourage the reader to start integrating data from the very beginning. From this perspective, reading can be seen as a continuous process of forming hypotheses, reinforcing them, developing them, modifying them, and sometimes replacing them by others or dropping them altogether.[30]

When you slow down the act of reading and scrutinize it, as we have just done with the prison/wrestling scenario, you cannot help but be struck by the sheer busyness involved. Reading even the most formulaic of genres can turn your mind into the equivalent of the *New York Times* newsroom, as you juggle convictions, conclusions, and predictions about everything in the text from the meaning of a word to a murderer's motives. Even moving from sentence to sentence is something of an acrobatic feat, particularly when we read passages like this one:

> Things were getting very tense. Suddenly John punched George and knocked him out. Mary started screaming. I ran to the phone and called the police. Kathy ran for the doctor.[31]

Under normal circumstances, most of us would assume that some dispute or suspected slight had been brewing between the two males in the passage before John decks George, and that Mary's screaming, Kathy's running for the doctor, and the narrator's actions are all related to the fisticuffs they have witnessed. Take a second look, however, and you will see whatever continuity or causality you assumed you had seen there evaporates under your gaze. Syntactically, there is nothing that tells us one event is following another, no adverbs, no phrases that establish a sequence. Nor do any of the sentences refer obviously to any other, no *seeing my distress, Kathy ran for the doctor*, no *since, if, because* formulas that could trigger our recognition of causation, nor any other flags that could prompt us to leap the synapse between sentences. Yet we do—constantly and unconsciously. Why?

First, these sentences seem both referentially coherent and plausibly related, largely because their contents correspond, more or less, to our experience of fistfights—from our days on elementary school playgrounds and hours of watching films or TV cop shows. It is relatively easy to infer that sentences 2 and 3 are causally linked, since Mary's screams would be a perfectly normal response to seeing two people she knows sock one another, or because sane people generally scream only in response to something upsetting they have perceived.[32] The remaining sentences appear to belong to the same scenario: running for the doctor and the police appear to be congruent actions when both occur in response to seeing a fight between two people who need either restraining or medical attention. As we will discover in the next chapter, the media we use for telling stories and conveying information all build off our proclivity to perceive the world in causal terms. We see cause and effect, and, unless discordant features rear their heads to disrupt our assumptions, most of us will leap with scarcely a pause—or conscious effort—from the sentence in which John punches George to Kathy's running for the doctor.

That is not necessarily true, though, of reading interactive narratives, particularly hypertext fiction, which can feel a little like a dimly remembered experience from Driver's Ed, perhaps your first attempt at piloting a car through heavy traffic. If you break down all the assumptions and calculations involved in, say, making a left turn, the sheer number of things you need to manage—monitoring the traffic on both sides of you, watching oncoming cars, keeping an eye out for erratic drivers and defiant pedestrians—seems so overwhelming, you can begin to be impressed with even the incompetent drivers in your family. In a similar fashion, reading hypertext fiction reminds us of just how complex the act of reading is, a condition to which a lifetime of immersion in a highly conventionalized, print-saturated environment has made us virtually immune. This is due, partly, to the absence of any established or even apparent conventions that guide the writers and readers who work with hypertext narratives. At the moment, the medium is an awkward grab-bag of conventions, practices, and techniques adopted from print that are not quite up to the demands posed by the technology. No prior or real-world models for paths or links currently exist, let alone conventions that tell us how to read or write cues that can help readers decide which paths to follow when they are lost or confused.

But we could also attribute this sense of reading as an almost overwhelmingly complex act to reading within an environment where the reader's convictions, predictions, and interpretations make a difference to the text itself. It is not simply a matter of hanging in there until you stumble across something you find vaguely comprehensible, as you can with an article or book, and then clutching on to your understanding of that fragment like a life preserver in heaving seas, hoping it floats you toward more manageable waters. Inside an interactive narrative, your understanding of a particular passage will determine what choices you make for moving on (or for shuffling backward a few paces, if you find yourself truly lost). If you are positive, after you put one murderer behind bars in *The Magic Death*, that the woman's brother or neighbor had equal cause to kill her, your convictions (no pun intended) can keep the story going. It can last, in this particular instance, until you are heartily sick of interviewing suspects and collecting samples of blood, fingerprints, and carpet fibers, are satisfied that you have nailed the true killer, or have run out of suspects (and murderers).

Because readers of interactive narratives can enjoy this newfound liberty to make choices and decide what deserves to become an "ending" to the stories they read, they also discover something that approximates Archimedes' fulcrum and level place to stand: a relative freedom that enables them to determine the satisfactions that closure is made of. Before this, closure was something we could describe and codify, but it was not something that we could examine outside of its role as a given. Although we have what seem like endless serials broadcast on television and radio (Britain's *The Archers,* for example, has been running for more than half a century) and horror films with deathless antagonists in the innumerable *Halloween* and *Nightmare on Elm Street* sequels, we have no existing models for narratives that lack tangible endings. Although you can break down the particular frustrations of an ending like that of *Jacob's Ladder,* which concludes by metamorphosing for one last time into a fantasy—of the "Oh, it *was* all a dream" ilk, familiar to us from *The Wizard of Oz*—it is a different task entirely, as we will discover, to see the ingredients of closure as they are defined by readers who can pick an ending, any ending—or none at all.

Examining how readers respond to interactive narratives, at its very least, enables us to nudge the act of reading and the pleasures readers take in it away from the long shadow cast by hundreds of years of readerly and writerly conventions. To do this, you need a more particularized understanding of just what interactivity is, an ant's-eye view of it that lets you grasp something that has no convenient, tangible likenesses, a virtual object that seems to stubbornly resist tidy analogies. Which leaves us—initially, in chapter 3—with a strategy that is a bit like a theoretical description of reading: a discussion of what interactive narratives are not.

What Interactive Narratives Do
That Print Narratives Cannot

I only wish I could write with both hands so as not to forget
one thing while I am saying another.
  —Saint Teresa, *The Complete Works of St. Teresa of Jesus*

If written language is itself relentlessly linear and sequential,[1] how
can hypertext be "nonsequential writing with reader-controlled
links," as Ted Nelson, who both created the concept and coined the
term, has argued?[2] How can we read or write nonsequentially, since
language, by definition, is sequential? Many definitions of hypertext
include this emphasis on nonsequentiality, as does the succinct
definition put forward by George Landow and Paul Delany in their
introduction to *Hypermedia and Literary Studies:* "Hypertext can be
composed, and read, nonsequentially; it is a variable structure, composed of blocks of text . . . and the electronic links that join them."[3]
But these definitions are slightly misleading, since both hypertext
fiction and digital narratives enable readers to experience their contents in a variety of sequences—as Nelson himself acknowledges in
Literary Machines.[4] As definitions go, those that emphasize nonsequentiality are also rather restrictive, since they tend to set hypertext
and hypermedia off from print in a kind of binary opposition: if print
is both linear and relentlessly sequential, it follows, then, that hypertext and hypermedia must be *non*linear and *non*sequential.

The dilemma in most short, succinct definitions of hypertext lies
in the definition of the word *sequence.* As used in the definitions
above, *sequence* and *sequential* denote a singular, fixed, continuous,
and authoritative order of reading and writing. But *sequence* can also
mean "a following of one thing after another; succession; arrangement; a related or continuous series," according to the likes of the
American Heritage Dictionary. In this context, it becomes significant
that the Latin root of sequence, *sequi,* means simply "to follow." All
interactive narratives have sequences—some of them more disorient-

ing than others, granted—making the medium, if anything, polyse-quential. The process of reading interactive narratives themselves is, as hypertext theorist John Slatin has noted, discontinuous, nonlinear, and often associative—but hardly nonsequential. His interpretation of hypertext accommodates Nelson's definition of "nonsequential writing" by inferring that Nelson meant "writing in that the logical connections between elements are primarily associative rather than syllogistic, as [they are] in conventional text," which closely corresponds to Bush's vision of the original Memex as well as the way in which most readers experience hypertext fiction.[5]

Arriving at brief and succinct definitions of an entire medium in a single sentence or even a mere phrase, at any rate, is more reductive than illuminating, a little like describing a book as "pages containing text that follows a fixed, linear order." While that might work perfectly well in describing instructions on how to operate your VCR, it doesn't quite cut it when it comes to nailing down the works of William Burroughs, nor does it account for the chapters on whales in *Moby-Dick* nor the likes of either *Hopscotch* or Barthes's *The Pleasure of the Text.* Moreover, it is not likely that anyone currently attempting to describe hypertext fiction, a medium that is only beginning to toddle through its infancy, is going to hit on an illuminating or time-resistant definition. Not only are the aesthetics and conventions of the medium evolving, but the technology itself is also still developing, as is its content, which currently borrows from genre and avant-garde print fiction, cinema, *Adventure* and arcade games, and graphic novels like *Maus.*

Further, as we have seen in chapter 1, critics, blinded by the small number of early works, have mistaken the hallmarks of a single type or genre of hypertext fiction for the defining characteristics of all present and future works within the medium.[6] This accounts partly for Birkerts's and Miller's flat rejections of hypertext fiction's aesthetic possibilities—although both critiques were probably also influenced by flawed assumptions about digital narratives threatening to replace print stories and novels. But this tendency to conflate early work and the aesthetic possibilities of the medium also sheds light on the puzzling critiques of hypertext fiction from otherwise insightful theorists like Janet Murray, who equates "literary hypertext" with postmodern narratives that refuse to "'privilege' any one order or reading or interpretive framework" and end up "privileging confusion itself."[7] If the earliest examples of hypertext fiction happen to represent the sophisticated play with chronology, completeness, and closure that draws many of its precedents from avant-garde print genres, it hardly follows that all hypertext fiction will resist privileging one reading of character or one set of choices for navigation through its network of potential narratives, or even that authors will plump for the conspicuously postmodern over, say, the hallmarks of the mystery, the hard-boiled detective story, or science fiction. Print fiction, after all, is hardly a monolithic entity: for every *Great Expectations* or *Persuasion* that Birkerts and Miller wish to defend from the onslaught of digital narratives, there are scores of Harlequin romances, John Grisham thrillers, and Danielle Steel paperbacks that their readers consume in a matter of hours and scarcely recall a week later. Print fiction means an abundance of genres and categories—*The Crying of Lot 49* existing alongside *Princess Daisy, The Bridges of Madison County* outselling *Middlemarch,* just as cinema includes both *The Magnificent Ambersons* and *Dumb and Dumber,* for all it may pain critics to admit it. This much is certain: the examples we have before us are only a beginning, the early efforts of writers who grew up with the singularity, linearity, and fixity of print. Imagine someone supplying an accurate definition of the content and aesthetic possibilities of all television programs once and for all during the Milton Berle era, when television borrowed heavily from vaudeville and theater, and you will have the right idea. For the purposes of investigating how readers experience and interpret interactive narratives in the here and now, it is far better for us to define just what hypertext fiction and digital narratives are and what they can do by examining just what they do that print does not—or cannot—do.

Interactive narratives have no singular, definitive beginnings and endings

"Begin at the beginning," the King said gravely, "then proceed straight through to the end. Then stop."
　　　　　　　　　　　　　　　　　　—Alice in Wonderland

Readers of print narratives generally begin reading where print begins on the first page of the book, story, or article and proceed straight through the text to the end. Although reading print narratives involves readers' thumbing back through the pages to clarify an impression or recall a name and a continual looking forward or predicting what will happen next, we nonetheless move more or less straightforwardly through *Pride and Prejudice* or *Huckleberry Finn.*[8] That is not to say that it is impossible to begin reading *The Great*

Gatsby at the point where Daisy and Gatsby are reunited for the first time in Nick's living room. But the reader who begins reading a print narrative in medias res is placed in a situation somewhat analogous to a filmgoer who has arrived in the darkened cinema forty minutes into a feature. Placed in these circumstances, we struggle merely to establish who is who and understand just what is taking place—and we bring to the text none of the opinions, expectations, conclusions, or, for that matter, pleasures that would otherwise be available to us had we followed the narrative from its beginning. The reader's gradual progression from beginning to end follows a carefully scripted route that ensures "the reader does indeed get from the beginning to the end in the way the writer wants him or her to get there."[9]

While many digital narratives begin with a scene or sequence that establishes both the identity of the user as part of an intrigue or quest and the parameters for the plot, most hypertext narratives have no single beginning. In Stuart Moulthrop's *Victory Garden*, readers are confronted with, among a multitude of possible ways of entering the hypertext, three lists that seem to represent a sort of table of contents: "Places to Be," "Paths to Explore," and "Paths to Deplore." Unlike a table of contents, however, these lists do not represent a hierarchical map of the narrative, providing readers with a preview of the topics they will explore during their reading and the order in which they will experience them.[10] The first place or path in the list has no priority over any of the others—readers will not necessarily encounter it first in the course of their reading, and need not encounter it at all. Each of the words or phrases, instead, acts as a contact point for readers entering the narrative. By choosing an intriguing word or particularly interesting phrase, even constructing a sentence out of a set of choices Moulthrop supplies, readers find themselves launched on one of the many paths through the text. In print narratives, reading the table of contents—if there is one—is generally irrelevant to our experience of the narrative itself: our reading experience begins with the first words of the narrative and is completed by the last words on the last page. In *Victory Garden* and most hypertext fiction, however, readers have to begin making choices about their interests and the directions in which they wish to pursue them right from square 1.

More strikingly, interactive narratives have no single "ending." *Victory Garden* has six different points of closure, while Michael Joyce's *afternoon* has five or more—depending on the order in which the reader explores the narrative space—since the sequence in which places are read determines whether or not readers can move beyond certain decision points in the narrative. And though the plot's puzzles, twists, and challenges in both *Gadget* and *Douglas Adams' Starship Titanic* culminate in a single endgame sequence that ratifies the reader's success in having solved the story's central puzzle, *Obsidian* challenges readers to allow the Conductor to live—resulting in the world as we know it being remade—or to destroy the Conductor and the Ceres Project and save the world. After making one last decision in *Obsidian,* readers still have opportunities to view the outcomes of the alternative scenario. More satisfying still are *Myst*'s three distinct endings that accompany readers' decisions to believe Achenar's, Sirrus's, or Atrus's version of events, and the eight potential endings to *Titanic: Adventure out of Time.* Deciding when the narrative has finished becomes a function of readers deciding when they have had enough, or of understanding the story as a structure that, as Jay Bolter notes, can "embrace contradictory outcomes."[11] Or, as one student reader of interactive narratives realized, as he completed a series of readings of *afternoon:*

> We have spent our whole lives reading stories for some kind of end, some sort of completion or goal that is reached by the characters in the story. . . . I realized this goal is not actually reached by the character, rather it is reached by our own selves. . . . [It] occurs when we have decided for ourselves that we can put down the story and be content with our interpretation of it. When we feel satisfied that we have gotten enough from the story, we are complete.[12]

This particular sense of an ending is, however, by no means unique to interactive narratives. Although print narratives physically end, literary conventions dictate that endings satisfy or in some way reply to expectations raised during the course of the narrative. As psycholinguists studying print stories have noted: "episodes end when the desired state of change occurs or clearly fails. In most stories, goals are satisfied and when goal satisfaction occurs, the protagonist engages in no further action."[13] In Stuart Moulthrop's interactive fantasy "Forking Paths," based on the Jorge Luis Borges short story "The Garden of Forking Paths," readers can experience no fewer than twelve separate instances of what we might call "points of closure"— places where the projected goals of the protagonist involved in a particular narrative strand are satisfied, or where the tensions or conflicts that have given rise to the narrative strand are resolved.

The multiplicity of narrative strands, the plethora of points of closure, the increased difficulty of reading interactive narratives—as we shall discover in the next chapters—can combine to stretch the time required to read an interactive novella like *Victory Garden*, with nearly a thousand segments of text and more than twenty-eight hundred links, to seventy hours. Compare this with the time required for the average reader to consume a three-hundred-page novel, generally anywhere from six to twelve hours.[14] Even a hypertext fiction as brief as Joyce's "Twelve Blue," with ninety-six segments of text bound by 269 links, contains multiple sequences that feed into other strands, crisscross them, loop endlessly, or arrive at points of closure, with no single reading exhausting the branching and combinatory possibilities of the text. Unlike print narratives, where each chapter builds upon the preceding one and leads to a single, determinate conclusion, the narrative strands in hypertexts can lead to numerous points of closure without satisfying the reader. Or the reader can be satisfied without reaching any point of closure at all.

Readers of interactive narratives can proceed only on the basis of choices they make

As noted in the previous chapter, in the past twenty years the concept of reading as a passive activity has become theoretically passé, an untenable stance held strictly by the unenlightened. Readers are now seen as breathing life into texts, reifying, or concretizing their possibilities—even receiving the text by creating it, in an effort nearly tantamount to that exerted by the author. As Barthes argues in "The Death of the Author,"

> [A] text is made of multiple writings, drawn from many cultures and entering into mutual relations of dialogue, parody, contestation, but there is one place where this multiplicity is focused and that place is the reader, not . . . the author. . . . [T]o give writing its future . . . the birth of the reader must be at the cost of the death of the Author.[15]

Yet reading print narratives is far from being a *literally* interactive activity, if we examine existing definitions of interactivity. Media theorist Andy Lippman of the Massachusetts Institute of Technology's Media Lab has succinctly defined *interactivity* as "mutual and simultaneous activity on the part of two participants, usually working toward some goal, but not necessarily"—a definition that can be met admirably thus far only by something as technologically unremarkable as human conversation.[16] For this "mutual and simultaneous activity" to be truly interactive, however, it must also, Lippman believes, contain a few other components.

Interruptibility: participants should be theoretically able to trade roles during the interaction, as speakers do in conversation, and not simply take turns in occupying the more active or more passive roles in the interaction.

Fine granularity: actors should not have to wait for the "end" of something to interact, with true interactivity being interruptible at the granularity level of a single word.

Graceful degradation: the parties involved can still continue the interaction without interruption, even if non sequiturs or unanswerable queries or requests enter into it.

Limited look-ahead: goals and outcomes in the interaction cannot be completely predetermined at the outset of the activity by either of the two parties, with the interaction created "on the fly," or coming into being only at the moment gestures, words, or actions are expressed.

Absence of a single, clear-cut default path or action: parties in the interaction cannot have definite recourse to a single or "default" path, one available to them throughout the interaction without their having to make any active decisions for interaction.

The impression of an infinite database: actors in an interaction need to be able to make decisions and take action from a wide range of seemingly endless possibilities.

When we converse, we stop or talk across each other (interruptibility)—often in the midst of a word or phrase (fine granularity)—and ask each other questions to which our partner may not have answers or even introduce non sequiturs into the conversation (graceful degradation). We can refuse to be cast in the role of cynic or idealist as we engage in an informal, conversational debate (no default), change subjects abruptly or follow an unforeseen shift in the direction of the conversation (limited look-ahead). Unless we find ourselves in the company of a true veteran bore, we seldom operate under the impression that our "database," the store of subjects and material from which we draw the shared opinions, emotions, and ideas that form the basis of the conversation, is anything but unlimited.

But according to this model of interaction, the average reader poring over *Jane Eyre* or *Ulysses* is placed in the position of someone listening to a monologue. We can interrupt only by closing the book or allowing our attention to wander, so the granularity to our interruption is the entire book itself. There is *only* one path through all but the most experimental of print narratives (these exceptions include *The Pleasure of the Text* or Julio Cortázar's *Hopscotch,* as we shall see). And if I try to focus only on the references to material wealth in *The Great Gatsby*—leaping from Daisy's voice sounding "like money" to a street vendor's absurd resemblance to John D. Rockefeller—my interaction with the novel will not simply degrade decidedly ungracefully, it will very likely collapse into mere incomprehension. My lookahead is also completely determinate and limited. If I become impatient with the unfolding of Agatha Christie's narrative *The Murder of Roger Ackroyd,* I can simply skip forward to the end and find out who bumped off Roger Ackroyd, and no matter where I pause to skip ahead—whether I stop at chapter 4 or 24, the murderer will always be the narrator. And, of course, my "database" will always be confined to the words in print enclosed between two covers, even if the significance of the text and the repertoire of interpretive strategies available to me were to embrace the entire existing literary canon.

Conversely, when readers open most interactive narratives, they can begin making decisions about where to move and what to read right from the outset, even, as in *Victory Garden,* right from the text's title. Most segments feature text that has individual words or phrases linked to other places or icons that act as navigational tools: arrows representing forward and backward movement, a feature of many hypertext narratives; the map of the United States and highway icons in "Trip"; a schematic map that recalls the London Underground journey planner and a map of the passengers in each car in *253;* the map of the ship in *Titanic;* a Mood Bar™ in *Midnight Stranger* that invites users to respond to characters by indicating green, amber, or red hues—presumably representing repartee that will push the conversation along, shift it into idle, or halt it in its tracks.[17] Unless segments are chained in a sequence with no options for navigation within each segment, readers can interrupt most interactive narratives within each segment—clicking on a word in *afternoon* or one of the brightly colored threads in "Twelve Blue," wandering up and down the seemingly endless corridors of *Titanic,* twisting doorknobs at random. The words, paths, and actions available as "interruptions," however, are chosen in advance by the author of the interactive narrative and not by the reader—an aspect of hypertext fiction that Espen Aarseth claims mitigates the medium's possibilities for bona fide interactivity, classifying it, instead, as we saw in chapter 1, as "participation, play, or even use."[18]

Furthermore, interactive narratives typically represent a spectrum of dialogues between reader and author anticipated in advance by the author, eliminating any possibility of graceful degradation. If I ponder the relationship between the unfaithful husbands and wives in *afternoon* and those in *WOE*, neither narrative can answer my query. Even the bots in *Starship Titanic*—ostensibly armed with thousands of lines of dialogue that should, at very least, enable them to respond to the words and sentences typed in by users—respond to lines within highly confined scripts. Insult Nobby, the elevator bot, and he cries, "Wot? Wot?"—the same response he'll also supply to a dozen other queries and statements. Pose a question to the snippy deskbot, and she replies tartly: "I'll ask the questions here," before proceeding with queries that you must answer according to a script; refuse to answer them or supply an answer different from those she obviously seeks, and you are doomed to listening to them repeated over and over again, ad nauseum. For all the developers of *Starship Titanic* may have labored for weeks over the bots' scripts, the main interaction remains between user and the tools necessary to defuse the bomb onboard the ship, replace Titania's head, and route the ship successfully home again, with the bots remaining intermediaries, obstacles, or helpmates in each of these tasks. And, contrary to Murray's belief that devices like *Midnight Stranger*'s Mood Bar™ make for less obtrusive interfaces for interaction, it can feel downright eerie to have a traveling businesswoman come on to you merely because you answered a seemingly innocuous query with a tap on the green end of the spectrum, particularly when you, the reader, are straight, female, and merely trying to locate the whereabouts of a mysterious intergalactic object.[19] Whenever interactions have been designed, the methods and consequences of interrupting them can feel more than a little limited or contrived.

Still, readers can meander around an interactive narrative in a manner not possible in print or cinema: in both *Titanic* narratives, I can wander around the transatlantic liner or the intergalactic spaceship at my leisure, examining objects, riding the elevators, making small talk with staff. As you amble around exploring, however, you eventually become aware that your actions have become decoupled from all aspects of the plot. Unlike a train jumping the tracks, however, your actions do not bring all potential for interaction with the text to a screeching halt. Your aimless explorations do, however, contain you within a temporal and plotless limbo, where time stands still and your interactions with bots, crew members, or passengers become severely restricted, if not impossible.[20]

While interactive narratives do not generally reward random explorations of the text—except when they happen to intersect with the plot's challenges and conundrums by pure chance—they offer readers a series of options for experiencing the plot, rather than the singular skein that connects print novels and stories.[21] On the no-default continuum, interactive narratives fall somewhere between the no-default absolute of conversation, where conversationalists may gamely try to answer you or listen even when you suddenly shunt the topic under discussion to something completely different, and the default-only mode of films—even on DVD or videodisc—where viewing segments of narrative in random orders makes a hash equally of plot, characters' motivations, causes, and effects.

In Web- and disk-based hypertext fiction, defaults generally take the form of arrow keys and represent the strongest links between one segment and another, usually tied together causally. In one scene in Carolyn Guyer's *Quibbling,* clicking on the Storyspace path key, an arrow, takes the reader from the segment where one character pries open a cigar box to the next segment in the sequence, where he hesitates as he opens it, and onto the next segment, where he peers inside. These links are called "defaults" in Storyspace terminology because they represent the action taken when readers choose to explore what may "come next," instead of choosing named paths to other segments from a menu or following links between segments connected by words in the text. Web-based hypertexts like "Trip" sometimes use default links to tie together narrative sequences that run to two or three segments so that readers experience and enjoy set pieces and vignettes as unbroken strings. Disk-based hypertexts, depending on the author's particular designs for potential interactions, may feature default links to and from virtually every segment of text, so that when readers reach the place "I call Lolly," in *afternoon* or "The End" in "I Have Said Nothing," the absence of a default can signal a potential ending of the narrative or a spot at which the readers must pause, reconnoiter, and decide whether—and how—to continue reading.

Even the presence of clear-cut links between causal sequences—or a single, clear-cut path through an entire narrative—does not provide a singular, authoritative version of the text that maintains priority over others. Defaults in *afternoon, WOE,* and *Victory Garden* do not provide a "master" version of the text.[22] Often, defaults deliberately play off readers' expectations, as in *WOE,* where readers using defaults shuttle between places describing passionate lovemaking between two couples. Because the default seems like the simplest and, therefore, most direct link between places, we assume that the stroking and groaning taking place between an unnamed couple in the first place we encounter belong to the same couple engaging in postcoital talk and smoking in the second. Since default connections do not involve us in the overt, more obtrusive acts of finding links in the text or choosing paths from a menu, hypertext readers may be tempted to see defaults as equivalents to the linear and singular connections characteristic of print. We discover this assumption with a jolt when we find that the couple in the first place consists of husband and wife, and, in the second, of the same husband and his wife's best friend. Default connections can jar readers, leap between narrative strands, and overturn predictions just as often as they can seamlessly move readers from one place to the next.

The impression of an unlimited database is not as impossible to convey as it may at first appear. The interactive narrative and simulation created by Mark Bernstein and Erin Sweeney, *The Election of 1912,* has 169 nodes containing information on the people, issues, and contexts surrounding the election, connected by an average of 4.3 links per node. Because this number of nodes can be comprehensively explored in one or two reading sessions, the database can seem conspicuously limited to readers. Yet, when these links and nodes are explored in the course of the decision making and planning involved in the simulated election of 1912—where readers manage Teddy Roosevelt's third-party campaign and enjoy a shot at changing history— the database seems considerably larger than a book of a comparable number of words. Because the information in each node appears in a dramatically different context, depending on the uses that the actor in the simulation finds for it, the database can appear to be double or triple its actual size.

The size of a database, the amount of information you have to potentially interact with, also depends on the number of pathways you can take through it. If you need to resort to the "back" option every time you want to explore more of a Web-based fiction, for example, your sense of the database can seem every bit as limited as it seems in *Gadget,* a highly atmospheric digital narrative that involves a comet hurtling toward the earth, a clutch of scientists creating retro machines straight out of *Brazil,* and a narrative that seems to lead almost inevitably to train stations regardless of the latest twist in plot. In *Gadget* the master narrative steers your experience ever forward, seamlessly, invisibly, through a world of train stations that recall the Gare du Nord and Waterloo Station—mammoth spaces that dwarf

football fields in which you sometimes discover your only navigational option involves strolling over to a phone booth where your alter
ego's detective superior, Slowslop, just happens to be waiting on the
other end of the line. If the train pauses at the station before your
assigned stop and you do not deign to step down from the car and stroll
over to a bystander on the platform—who, not coincidentally, has a
tidbit of information about the scientist you are stalking—the narrative stops dead. Most digital narratives are built around a quest,
whether for the identity of a killer or artifacts collected on a grown-up
version of a treasure hunt, providing a set of purposes that inform the
narrative, propelling both it and the reader forward. The quest also,
conveniently and not merely incidentally, enables designers to limit
the characters, spaces, and scenarios populating the narrative.

Grail-less *Gadget*, which requires its readers merely keep going
through the narrative, is, however, more immersive than *Myst* or
Obsidian because its readers seldom need to pause and think purpo-
sively about the plot, plan some strategic swordplay, or collect the
obligatory artifacts that litter so many digital narratives. Ironically,
Gadget derives its ability to lure readers into the externally oblivious,
trancelike state of ludic reading precisely because its database is
severely limited: you do not need to poke around the hotel for a map
that will let you locate the train station. In fact, if you do not pause for
a word with the clerk hovering over the reception desk, you cannot
leave the hotel, let alone get to the train station, because the clerk conveniently has your ticket. Pick up the ticket, and the entire scene dissolves gently to the train station, segueing to the spot where a ticket
agent retrieves the ticket from you. Likewise, if you attempt to leave
the cavernous Museum train station without a second conversation
with the distinctly odd-looking character lingering by the steps, your
cursor will not turn into the directional arrow enabling you to navigate down the stairs and out of the building. Occasionally, the participatory and immersive aspects of interactivity can become mutually
exclusive, one reason a narrative with a small database and virtually
illusory choices for navigation should, nonetheless, seem peculiarly
compelling, even entrancing.

What is striking about narratives like *Gadget* is that too much
participation, too many gadgets to collect and assignations to keep and
bad guys to sock, detracts from the immersiveness of digital environments, the very feature that Murray believes represents their single
most valuable aspect. Constant demands for input or inputs that are
frustrated—as when, for example, players thrash around *Myst*'s land-
scape, clicking wildly and randomly in the fervent hope the shape of

their cursors will change and permit them to move forward in the narrative—can remind readers that they are grappling with a narrative designed by others, disrupting their suspension of disbelief in the same way that difficult texts do: requiring frequent pauses, reflection, even regressing over pages already read.

Paradoxically, genre fiction and interactive narratives like *Gadget* that are not terribly interactive fill readers' cognitive capacities more completely than difficult texts; familiar plot conventions and characters considerably speed the pace of reading and absorption, placing a far heavier continuous load on readers' attention.[23] Authors may use default options to privilege some linkings in the text over others, saving the fates, for example, of characters until readers know them sufficiently well for their victories or deaths to matter. Authors may use defaults to remove readers' concerns about actions and paths taken, thereby deepening their immersion in the narrative. And sometimes they use defaults to limit the amount of sheer data any interactive narrative must include to produce even a small simulacrum of a mere wedge of the world.

Finally, interactive narratives offer a very tangible sense of limited look-ahead, because navigational choices always depend on where you are and where you have already been. Occasionally, since connections between places can crisscross each other in a truly tangled skein, readers attempting to re-create an earlier reading exactly, by using, say, the "back" option on their Web browser, can find it well-nigh impossible without following a list of their previous navigational choices. You cannot be entirely certain, either, that your carefully considered choice has not triggered a connection randomly—as it can in Storyspace narratives when the author creates more than one default branching out from a single place—so that the same answer to the same question does not yield the same reply. This makes your reading of hypertext fiction a far less predictable matter than conversation with most people, even those you know only slenderly, since most of us exchange words according to highly structured conventions that extend from gripes about the weather to a confession of the strangers-on-a-train variety made aboard the Twentieth Century Limited. That means that, while our look-ahead in conversation is limited—even if I have already agreed with my partner not to mention the Clintons, the stock market, or whether the Rolling Stones should throw in the towel and retire—I also cannot begin to see what is coming next when, for example, you start talking about the War of Jenkins' Ear. When I read *afternoon,* though, I have no way of knowing where the narrative may branch next, where any of the connections I choose may take me, or how long my reading of the text will take (which can last as long as my eyes hold up). As far as the limited look-ahead corollary goes, where interactive narratives are concerned, you *can* have too much of a good thing.

On the other hand, some hypertext fiction provides readers with the kind of overview impossible in a face-to-face exchange, via functions like the cognitive maps in Storyspace that act as schematic drawings of all possible versions of the text you might experience if you persevere long enough. As I pursue a narrative strand in *WOE* concerning the couplings and uncouplings of the adulterous foursome, I discover that all are connected by a single path named "Relic" and that, by selecting "Relic" from the Path menu each time it appears, I can watch the four come together in various combinations throughout their daily lives. When I encounter the place called "We," I stumble across a concluding sentence that reads "a happy ending," something that seems entirely at odds with the heavy sense of foreboding that seems to hang over the characters. When my desperate search for any further places on the "Relic" strand proves fruitless and subsequent browsing through a succession of nodes yields no further trace of the "Relic" foursome, I quickly switch to the Storyspace cognitive map and find "Relic" at last: a chain of places tidily laid out within a single, confining space and connected by path arrows labeled "a story" that ends with the place "We." The "happy ending," it turns out in this version, really was an ending, which makes me reconsider if the adjective, then, should be read ironically after all—an interpretation possible only through my using the map of the text to gain an Olympian perspective over the entire thing, what Jay Bolter has called a "structure of possible structures."[24] Like a topographic map of an unfamiliar island, the cognitive map of *WOE* eases the limitations of my look-ahead, providing me with vague suggestions about which directions might prove the most fruitful for further, dedicated exploration.

Interactive narrative segments exist in a network of interconnections mapped in virtual, three-dimensional space

It is not necessary to pore over cognitive maps, or any map at all, to encounter interactive narratives as structures suspended in virtual, three-dimensional space. In a look at the interpretive strategies used by readers of his own "Forking Paths," Moulthrop discovered that maps are not essential to navigation through hypertext space but that readers of hypertext fiction seldom read without an awareness of the virtual, three-dimensional arrangement of the places they read. Back when he was still adapting the hypertext to the fledgling Storyspace interface, Moulthrop casually provided me a copy of "Forking Paths"—which I took straight into a freshmen expository writing workshop I was then teaching at New York University. Dividing the class in half, I asked students to retell what they thought happened in the texts they read, then handed out photocopies of Borges's short story "The Garden of Forking Paths," and diskettes with copies of Moulthrop's hypertext. Still unpublished, "Forking Paths" is a hypertext fantasy built around a skeletal arrangement of the Borges short story, with fully fledged narratives branching off from each of the episodes and scenarios depicted in the original print fiction. Intending to invite readers to become coauthors of "Forking Paths," Moulthrop had omitted default connections and relied entirely on links, joining places through words or phrases in the text of each place. This, he explains, seemed logical to him, because "stories are a dialectic of continuity and closure, each fragmentary unit of the text (word, sentence, page, scene) yielding to the next in a chain of substitutions or metonymies that builds toward a final realization of the narrative as a whole, or a metaphor."[25] Although he acknowledges that the readers may have been somewhat disabled by the lack of instructions (which were still being written for "Forking Paths"), when he read their written responses to his hypertext he discovered the antithesis of what he had anticipated. Instead of engaging the text at the local level and reaching what critic Peter Brooks has described as a metaphor for the text through following a chain of metonymies, my students gave up attempting to discover matches between their choices of words to form likely links between places and the words Moulthrop used to link them.[26] Amid all the complaints, however, one enterprising reader hit on navigation buttons that enabled him to move up, down, left, or right from the place he was stuck in. Others followed suit, exploring the hypertext outside the connections Moulthrop had mapped for them. As a result, their discussions of the narrative strands and the narrative as a structural whole reflected their awareness of moving through this virtual space, much as Greek and Roman rhetoricians once mentally strolled through their elaborate memory palaces. Inverting the relationship between metonymy and metaphor implicit in conventional print narratives, my students

were plotting their own readings through a cartographic space, hoping to discover a design which, though it was in no way "promised," might prove to be buried or scattered in the text. The map, which represents the text as totality or metaphor, was not something to be reached through the devious paths of discursive metonymy, rather it was a primary conceptual framework, providing the essential categories of "right," "left," "up," and "down" by which these readers oriented themselves.[27]

As Jay Bolter argues, "topographic" writers in print—Sterne, Joyce, Borges, and Cortázar, who have created narratives that explore, exploit, and chafe at the confines of printed space—are "difficult" writers.[28] What makes them difficult is their self-conscious absorption with the act of writing itself and the problematic relationship among narrator, text, and reader, since their print texts work strenuously— and ultimately unsuccessfully—against the medium in which they were conceived. This is largely because spatial relations in print narratives—or the "spatial form" lauded by Joseph Frank and his critical successors—are very much like spatial relations in the cinema, where we see three dimensions represented and projected on a flat, two-dimensional plane.[29] We understand that the placement of the objects, characters, and events represented in print narratives has significance in terms of our understanding of the entire work, but this understanding is not necessary to our ability to proceed through the text itself (although, upon seeing his first film, an actor once reported, he and the other children watching it in the humid island cinema ran out into the alleyway behind the screen in search of the police car that had raced from one side of the screen to the other). Our awareness of print space, containing two potential dimensions, and of cinema, three dimensions projected onto two, is intrinsic to our reading experiences of both media.[30] In hypertext narratives, however, this awareness is inextricably wedded to our "reading" of the text itself, because the burden of interactivity and the continual necessity to choose directions for movement never allows us to forget that we are reading by navigating through virtual, three-dimensional space.

Interactive narratives have many orders in which they can be read coherently

As Richard Lanham has observed, digital media—such as digitized films and interactive narratives—have no "final cut."[31] This means they have no singular, definitive beginnings, middles, or endings, but also that no single, definite order of reading is given priority over the others that exist alongside it. There is also no single story, and, contrary to our expectations based on reading print narratives, readings do not simply provide varying versions of this story or collection of stories. As Jay Bolter has argued, each reading generates or determines the story as it proceeds: "[T]here is no story at all; there are only readings. . . . [T]he story is the sum of all its readings. . . . Each reading is a different turning within a universe of paths set up by the author."[32]

In *afternoon,* some readings represent alternative voices or perspectives on the narrative, with the changes in narrative perspective made separate and discrete by electronic space. The narrative strands in *Victory Garden* involve political developments during the Nixon, Reagan, and Bush eras, paralleling and crisscrossing each other as they follow a few weeks in the lives of nine characters. In "Twelve Blue," narrative strands represent the perspectives and experience of each character, each strand corresponding to the brightly colored threads that cross, arch, and dip across a blue field, a visual corollary to the voices and stories contained in the narrative that touch each other when stories meet or fray at the ends as stories begin to wind down. In some instances, the readings themselves may constitute mutually exclusive representations of the same set of circumstances with radically different outcomes, as readers can discover in both *afternoon* and "I Have Said Nothing." Like these hypertext writers, Faulkner once attempted in print to separate the different perspectives in *The Sound and the Fury* with something more than the conventional options of white space or discrete chapters. When, however, he indicated to his publisher that he wanted each represented by different colors of ink, Random House shuddered at the cost and refused.[33]

When you read hypertext narratives, you also have the option of limiting your experience of the text to the pursuit of narrative strands that you find particularly intriguing. If I want to trot after the romance burgeoning between Nick Carraway and Jordan Baker in *The Great Gatsby,* I have to read, or browse through, or skim the entire novel in order to pursue the romance that mirrors Gatsby's involvement with Daisy. And, of course, this narrative strand, like the episode narrated by Jordan, is but a fragment of the total novel—a particle that is comprehensible and meaningful only in the context of the novel as a whole. Yet I can simply pursue the tortuous relationships between the unfaithful wives and husbands of *WOE* or focus my readings on the relationships between Emily, Victor, and Jude as I make my way through *Victory Garden.* In some instances, focusing on the stories and strands of particular interest may be relatively easy, with the options for navigation through the narrative made accessible through lists, as in *Victory Garden*, or by way of cognitive maps that enable readers to arrive at a place by pointing at it with a cursor. At other times, however, following a single narrative strand can involve a complicated process of selecting paths by trial and error, or determining which path or place names document certain narrative episodes and strands. Regardless of whether the process of following the chosen narrative strand is easy or incredibly difficult, readers of hypertext narratives can coherently experience these texts in a variety of different orders and sequences without doing violence to the narratives, stories, or meaning of the hypertext as a whole.

The language in interactive narratives appears less determinate than the language present in print pages

Most obviously, interactive narratives embrace a far wider and less determinate spectrum of meanings than print narratives because few readers will experience identical readings of texts that can have thousands of connections between thousands of segments of text, which can be as brief as a single word or as long as pages of print text. The more links, or decision-points, each reader must confront in the course of navigating through the narrative, the less singular and determinate the meaning of the hypertext narrative as a whole, because no single path through the text has priority over all others.

Yet the indeterminacy of interactive narratives also exists in a far more fundamental sense than this. In most hypertexts, a majority of the nodes will appear in more than one context as a point along two, three, or more paths. The metaphor for hypertext is, after all, not a flowchart but a web that acknowledges the myriad of associative, syllogistic, sequential, and metatextual connections between words, phrases, paragraphs, and episodes. To be comprehensible, print paragraphs need only to build off the paragraphs that have preceded them and prepare the reader for what is yet to come. Print narratives can use paragraphs and transitions toward creating a sequence that both directs the reader's experience of the material forward and seems like the most authoritative, and even the only possible, sequence for structuring the material.[34] But hypertext fiction seems to work in the opposite direction. Ideally, print paragraphs and transitions close off alternative directions and work to eliminate any suggestion of other potential sequences that might have been created from the same material—so that readers do not end up stopping in the middle of a paragraph like this one to reflect on all the other ways these same details might be construed. But nodes or windows in hypertext fiction must, by their very nature, prove comprehensible in more than one sequence or order. Instead of closing off any suggestion of alternative orders or perspectives, the text contained in each segment must appear sufficiently open-ended to provide links to other segments in the narrative. This, de facto, fosters an additional level of indeterminacy generally rare in print narratives—although it does appear in avant-garde and experimental forms of print narratives like *The Alexandria Quartet, Hopscotch* and *The Pleasure of the Text.*

Print Precursors and Hypertext Fiction

At present, existing hypertext fiction resembles two of the divergent modes explored in avant-garde or experimental fiction: what we might call "narratives of multiplicity" and "mosaic narratives." Mosaic print narratives, such as Lawrence Durrell's *Alexandria Quartet*, Julio Cortázar's *Hopscotch*, and Barthes's *Pleasure of the Text* consist of narrative fragments, conflicting perspectives, interruptions, and ellipses that impel readers to painstakingly piece together a sense of the narrative, with its full meaning apparent only when viewed as an assembled mosaic, a structure embracing all its fragments.

At a local level, a mosaic narrative such as *The Alexandria Quartet* presents its readers with more determinacy than *The Pleasure of the Text.* That is, Durrell's novel consists of a set of four novels each of which can stand as a discrete, independent text on its own, and each seems perfectly conventional and self-contained when read separately. Unlike trilogies or tetraologies that merely feature the same bit of geographic territory or the same cast of characters, *The Alexandria Quartet* novels relate the same set of events from the perspective of the different players involved. Even readers of *Justine*, the version of events narrated by the naive Darley, can feel their experience of the novel is perfectly complete when they reach the ending. Yet, as you move from *Justine* to the last of the novels, *Clea*, your view of events begins to burrow beneath the skin of the world according to Darley and the worlds known by Balthazar, Mountolive, and Clea, the most informed of the four narrators. By the time you reach the end of *Clea*, the observations made by Darley in *Justine* that had seemed so straightforward and reliable can end up seeming a little like Benjy's in *The Sound and the Fury*. What had appeared perfectly accurate in even *Balthazar* and *Mountolive*, when read against Clea's supplementary version of events, brims with ambiguities, ellipses, and unanswered questions, making you wonder how you had ever accepted it as a fully fledged account in the first place. Balthazar's story points up how hopelessly uninformed Darley's grasp of reality is and positions itself as an authoritative supplement to it. Balthazar's representation of Nessim's proposal to Justine, meant to provide us with insight into their relationship, insists that Nessim is hopelessly infatuated and Justine ruthlessly pragmatic:

> After a long moment of thought, he picked up the polished telephone and dialled Capodistria's number. "Da Capo," he said quietly. "You remember my plans for marrying Justine? All is well." He replaced the receiver slowly, as if it weighed a ton, and sat staring at his own reflection in the polished desk.[35]

Hundreds of pages later, in *Mountolive,* you may find yourself wondering just how penetrating Balthazar's insight was when you encounter the same scene again:

> [A]fter a long moment of thought, [he] picked up the polished telephone and dialled Capodistria's number. "Da Capo," he said quietly, "you remember my plans for marrying Justine? All is well. We have a new ally. I want you to be the first to announce it to the committee. I think now they will show no more reservation about my not being a Jew—since I am to be married to one."[36]

Plainly, Balthazar's story about the personal relationship between Nessim and Justine cannot do full justice to the complexity of their passionate political and strategic alliance, and our understanding of the entire world of *The Alexandria Quartet* shifts dramatically from the inclusion of a mere three sentences. Our faith in the accuracy and authenticity of Balthazar's account, which presented itself as more complete than Darley's, is tattered well before the close of *Mountolive,* just as the value of the Mountolive section declines seriously the further we proceed through Clea's. You could not, however, save yourself the effort of reading all four novels simply by beginning with Clea's account—that would be rather like chipping a diamond apart so you could admire the slender sliver of its face and lose the pleasure of peering beyond it into depths emphasized by precisely cut facets.

The pleasure of reading Durrell's tetralogy is not unlike the pleasure in listening to Bach's *Goldberg Variations,* where you are dazzled by just how richly evocative a few seemingly simple phrases can be—

here sequence is everything. In Barthes's *Pleasure of the Text*, sequence apparently means nothing: the book itself is a succession of fragments, ordered alphabetically. While the segments are tagged with titles in the book's table of contents, the reader in the throes of absorbing the text has no such assistance, only a scattering of typographical marks and white space to indicate the division between fragments. Together these pieces represent Barthes's erotics of the text, yet no single fragment maintains priority over the others, and even the most vigilant readers will not find any transitions to transport them easily and painlessly into the next segment. As Barthes notes in one such segment, "[A]ll the logical small change is in the interstices. . . . [T]he narrative is dismantled yet the story is still readable."[37]

The Pleasure of the Text offers the same lack of definitive beginnings, middles, and endings, and singular, definitive paths through the narrative you would discover in hypertext narratives. Likewise, Durrell's *Alexandria Quartet* presents readers with the discrete, separate, and entirely self-contained narrative perspectives that you could encounter in the likes of Moulthrop's *Victory Garden* or Joyce's *WOE* or *afternoon*. Yet each of Barthes's segments and Durrell's chapters builds off the others in a highly determinate way impossible in hypertext fiction. Read in a random, reverse-alphabetic order, Barthes's meditations on the act of reading do not bear upon one another any differently than they might if you were to explore the text from front to back, or to weave your own path through the book. If there are alternative ways of assembling Barthes's erotics of text, other orders awaiting liberation from the linearity of conventional print, they do not crowd the surface of the text or shout at you from its pages, which are, after all, still relentlessly linear. Similarly, Durrell's presentation of four sequential narratives trace and retrace the same events in a chronological order that removes any ambiguity from your immediate experience of the narrative. As you ponder the entire construction in retrospect in light of what you have learned by the end of *Clea*, what is striking is not how ambiguous or incomplete events seem (since the version presented in *Clea* fills in any last vestiges of ambiguity or openness) but how obtuse or slender a grasp any of the observers have on the complexity of the whole. At no point in the throes of peering over Darley's shoulder, though, or reading Balthazar's notes, are you invited to mull over what might be missing from their depictions of events: ambiguity here is something you are free to realize had been present only after a fully informed, detailed account has banished it forever.

Just as you are not aware, the first time you happen upon Nessim's telephone conversation, that you are not getting the whole picture (nor that you are going to see it replayed again somewhat differently), you probably would not find one particular passage in *afternoon* remarkable the first time you run across it. In it, the protagonist and sometime narrator, Peter, shares lunch with his employer, Wert. There is a bit of badinage, some sexual innuendo reserved for the waitress, and then Wert springs a question on Peter:

> He asks slowly, savoring the question, dragging it out devilishly, meeting my eyes.
> <How . . . would you feel if I slept with your ex-wife?>
> It is foolish. She detests young men.[38]

The second time you read this, however, you might be convinced that you had read a different passage, and, by the third or fourth time, you might find yourself trying desperately to locate these different spots that sound awfully similar but seem to mean entirely different things. In one narrative strand, this segment crops up amid Wert's clowning around over lunch, emphasizing his immaturity around women. In another, Wert poses the question to Peter playfully, to distract him from his concern over the whereabouts of his missing son and estranged wife, whom he believes may have been injured in a car accident earlier that day. Encountered in yet another context, the passage occurs in the context of Peter's fling with a fellow employee, Nausicaa, and Peter sees Wert's question as evidence of his boss's jealousy over their involvement. Later, the lunch date and conversation reappear after a narrative strand couched in Nausicaa's own perspective, which reveals that she is sleeping with *both* Wert and Peter, making Wert's query something of a game of cat's-paw. "I'm sleeping with your lover," Wert seems to be thinking, so he follows the line of thought to a position he perceives as more daring: "What if I were sleeping with your ex-wife?" But if you reach a segment called "white afternoon," having visited a fairly detailed series of places, you will discover that Wert and Peter's ex-wife, Lisa, have been seen together by Peter himself, although Peter cannot be certain that they are involved with each other. When the lunch time conversation reappears, after this last revelation, Wert's query is a very real question indeed.

What is striking about the way *afternoon* works is that there is only one passage involved here—and the language within it is as fixed as on any printed page. Although the contexts may alter its meaning drastically with each new appearance, the language itself stays the same, unlike Durrell's quartet of novels, where he can manipulate our perspective on events only by a combination of ellipsis and supplement. Yet the language itself is not indeterminate: readers seeking a precedent for the "he," "my," and "she" that occur in the text need look no further than the preceding or succeeding segments. In all the contexts in which this place appears, it is clear that the "he" posing the question is Wert, the ex-wife or "she" in question is Lisa, and the "you" who thinks the question is foolish is Peter.

In *WOE* Joyce further capitalizes on the indeterminacy of hypertext narratives to induce a reading experience that approximates a trompe l'oeil, where your interpretation of what is happening in a narrative sequence disintegrates just as you finish reading it. It would not normally occur to you to wonder if the "he" you have been reading about is the self-same "he" a few paragraphs later, but *WOE* springs its surprises on you by switching the identities of pronoun precedents in midstream. You can never be certain who the "he" and "she" are in a particular passage—to brilliant effect, since several of the narrative strands in *WOE* involve romantic attachments between two couples closely allied by both friendship and infidelity.

The other form of print narrative that thrives on indeterminacy, the narrative of multiplicity, is produced by writers who have chafed at the way confines of printed space preclude multiple, mutually exclusive representations of a single set of events. Robert Coover's "Babysitter," and "The Elevator" from *Pricksongs and Descants,* Borges's "The Garden of Forking Paths," and Fowles's *French Lieutenant's Woman,* all engaging and entirely successful works of fiction when read at face value, are also as much about the experience of multiplicity and simultaneity and the way these are represented in print as they are about their ostensible subjects.

Fowles's *French Lieutenant's Woman,* for example, features three endings: a parody of the tidy-but-breathless tying up of loose ends so characteristic of the Victorian novel; a happy but conventional resolution of the tortured relationship between Charles and Sarah; and a more complex, "modern" resolution that serves to deconstruct the paternalistic perspective of the traditional Victorian novel of love and marriage. Not surprisingly, none of the three endings is compatible with another. Tellingly, the modern, deconstructive episode comes last in print—which can be said to provide this last "ending" with priority over those preceding it—just as the "ending" that occurs midway through the book has its authoritativeness somewhat undermined by the bulging stack of unread pages remaining after it.

More radically, Coover's "The Babysitter" features 105 narrative segments that begin as nine separate and distinct narrative strands framed from nine different perspectives, becoming less distinguishable from one another as the narrative proceeds. Mutually exclusive versions of events begin unfolding one after the other, sometimes feeding clearly into each other. The passages depicting husband Harry's first sexual musings on the babysitter and wife Dolly's bitter thoughts about marriage occur sequentially in time, united by Dolly's question, "What do you think of our babysitter?" which appears in both segments. By the time the reader has reached a section where the babysitter screams after discovering herself watched from a window, however, it is not clear whether the perspective originates in her boyfriend's fantasies about her or in Harry's idylls of seducing her. In the segment that immediately follows it, the babysitter's scream metamorphoses into an indignant shriek as the children she is supposed to be supervising whisk the bath towel away from her wet body after she leaves the bathtub to answer a phone call. The phrase "she screams" is identical in both passages, but the context and narrative strands in which it is embedded are mutually exclusive representations of a single moment in time. In the narrative universe of "The Babysitter," all possibilities are realized, with actions, thoughts, idylls, and snatches of television programs offering an equal, textual tangibility. In the end, however, all this burgeoning and splintering of perspectives converges in two episodes. One neatly resolves the wild and mutually exclusive seduction, rape, and murder scenes by depicting the babysitter waking up from a dream amid a setting so orderly that even the Tucker family dishes have been washed and put away. The other represents a conflation of all the narrative strands in a single, final, wild conclusion: the Tucker children are dead; the babysitter is a drowned corpse in the bathtub; Mr. Tucker has fled the scene; and Dolly cannot get out of her girdle.[39] The wild improbability and satirical tone of the last segment and the suggestion, in the passage that precedes it, that everything in the narrative belonged to one vast, distended dream, also tends to undermine the "reality" and priority of any single narrative segment or sequence.

When print narratives attempt to resist the physicality of print by increasing the number of stories, narrative strands, and potential points of closure—as is the case with the likes of "The Babysitter," Sterne's *Tristram Shandy*, or Borges's "The Garden of Forking Paths"—the medium inevitably resists, making the reading experience and the significance of the narrative itself a meditation on the confines of print space. In "The Garden of Forking Paths," for example, master spy Yu Tsun is introduced by Sinologist Stephen Albert to the concept of the labyrinth, once discovered by his illustrious grandfather Ts'ui Pên. The embodiment of an "infinite series of times . . . a network of diverging, converging and parallel times . . . [that] embraces *every* possibility,"[40] the labyrinth represents an alternative universe where mutually exclusive possibilities exist alongside one another, producing a space in which, as Albert himself notes, Yu and Albert can simultaneously be both friends and mortal enemies. Yu is, of course, both to Albert. As the grandson of Ts'ui Pên, he shares Albert's most valued interests; as a German spy who must kill Albert to signal the location of a British armaments site in France, he is also his most deadly enemy. With the arrival of the British captain pursuing Yu, however, the German spy shoots Albert and the infinite possibilities hinted at in the story are, ironically, reduced to a single, sordid conclusion—death.

Seeing the story as an example of topographic writing struggling against the confines of print, Stuart Moulthrop sought to liberate the Borges story by splitting the original story into hypertext nodes, then grafting onto each node a series of narrative strands.[41] Following certain links introduces you to a narrative involving Stephen Albert's former lover or thrusts you onto the trail of Yu's German intelligence chief, Viktor Runeberg. You might follow Yu, Captain Richard Madden, and Albert through the labyrinth in the garden and experience no fewer than twelve separate permutations on the ending to the original Borges tale. Or you might explore retellings of narrative events plucked from the original story from entirely new points of view, unexpected reversals in character traits and motives, and even playful, metatextual commentary on interactivity itself. In Moulthrop's garden, Yu Tsun murders Stephen Albert; Albert and Yu stroll peaceably into the labyrinth together; Yu disappears from pursuer Richard Madden in the midst of the labyrinth; and Albert garrotes Yu—a true realization of the "infinite but limited" labyrinth of possibilities that exist in the heap of contradictory drafts that constituted the Garden of Forking Paths created by Ts'ui Pên.

Similarly, in Michael Joyce's *afternoon*, car accidents occur, seem to have occurred, may possibly have occurred, and never occurred. The narrator, Peter, has an affair with Nausicaa but also does not have an affair. His employer, Wert, is faithful to his wife, is having an affair with Nausicaa, and may well have had an affair with Peter's ex-wife, Lisa—or none of the above. Peter loses his son, fears him dead or seriously injured, and begins a frenetic search for him in some readings of *afternoon*. In others, he simply goes about his daily business. "The story," Jay Bolter has noted, "does and does not end."[42] There is a challenge implicit in any reading of these highly indeterminate narratives that embody a dense thicket of possibilities without giving priority to any one of them, a requirement that we learn to read "multiply," as Bolter insists (144), aware that a single perspective on any set of circumstances can never do full justice to the complexity and contingency of even a fictional world dreamed up by a single author.

All right, you may be wondering, so interactive narratives do not have singular or definite beginnings or endings, and readers can proceed through them only by making choices about what they have read or what they would like to read . . . but how on earth do you know when the story is over? How do you know when it is finished, when you are finished? Most of us have, at one time or another, flinched at the credits scrolling up the screen, wondering how the story could be over when so many loose ends were left dangling so teasingly. We are accustomed to dealing with texts that end more prematurely than their stories would seem to, but what do we do with a text that, a bit like a book made of sand, has pages we cannot properly count and nothing like end titles or hard covers to contain it? And, when you stop reading, what is really finished: the stories—or you?

Charting Maps and Raising the Dead:
Readers' Encounters with Hypertext Fiction

[In discourse] all the logical small change is in the interstices.
—Roland Barthes, *The Pleasure of the Text* (1975)

The essence of cinema, it has been suggested, is the cut.[1] What is a film, after all, other than a succession of strips of celluloid, a series of various shots linked together with editors' tape and shown as one continuous strip? As we sit in darkened cinemas watching features by the likes of Hitchcock and Sirk—or even Coppola and Scorsese—we no longer see the cuts or the edits: we see the connections. If we see a shot of an unshaven man half-heartedly slurping soup in a grimy diner, followed by a shot of an impeccably groomed croupier presiding over a craps table in a crowded casino, followed by a shot of an elderly woman lying amid a forest of tubes, drains, and drips in an intensive-care unit, we do not perceive a sequence of random shots or images. Instead, we see this sequence as either a piece of scene setting, establishing locations that will later play critical roles in the narrative as it unfolds, or as an introduction to the narrative's dramatis personae. We also tend to assume that these shots are joined in roughly chronological time, that we are dipping into these characters' lives at approximately the same time, and that the coupling of their lives in this short sequence of shots anticipates real interactions and encounters to be revealed during the course of the film.

Are these assumptions learned, our expectations the natural outcome of our being thoroughly schooled in the grammar of cinema, or are they endemic to our way of perceiving the world? In the fledgling Soviet Union, due to an acute shortage of film stock, pioneer film theorist Lev Kuleshov found himself able to create films only by slicing and splicing existing film footage from wildly different sources. Using a single, lengthy take of an actor's face with a neutral expression, Kuleshov sliced it into three equal segments and intercut these with images of a bowl of steaming food, a dressed out corpse, and a child playing. When he projected his completed "film" to an audience barely conversant with the new medium, however, they praised the actor's subtle ability to convincingly portray the emotions of hunger, grief, and joy.[2] Our ability to perceive connections such as these, clearly, has little to do with our knowledge of the grammar of shot–reaction shot (which dictates that each shot following the close-up of the actor's face reveal the objects he or she sees) and everything to do with the nature of human perception.

Narrative, Jerome Bruner has argued, can be defined by a single, serviceable criterion: narrative is that which deals with the "vicissitudes of intention." The reason for the role of intention in defining narrative is clear, he continues: "[I]ntention is immediately and intuitively recognizable: it seems to require for its recognition no complex or sophisticated interpretive act on the part of the beholder."[3]

Narratives, in a sense, are about connectedness, sequence, order—qualities that are inextricably linked to the way we view the world around us. We might even say that narratives represent a reflection of our tendency to perceive the world in terms of intentional or causal states, albeit a reflection that produces an orderly, predictable, and complete world within a static structure. But if this mode of perception is part of what makes narratives so attractive to us and, apparently, so important a component of our lives, it is also integral to our being able to read, interpret, and understand narratives at all.

Most theories of reading have ignored the biases of human perception toward "seeing" causality and intention even where these qualities may not exist. In a study of perception conducted in Belgium in the early 1960s, subjects asked to watch projected images of moving, animated objects persistently saw causation connecting movements. Invariably, subjects discussed the movements they had watched in strictly causal terms, with the objects perceived as "dragging," or "deflecting" one another. Although the subjects were conversant with physical laws that could explain the movement of the objects they saw, the presentation of two or more objects invariably prompted the subjects to see their movements as strictly "caused."[4] In a similar experiment, psychologists Fritz Heider and Marianne Simmel also projected an animated film featuring a small moving triangle, a minute moving circle, a large moving square, and an empty rectangle to viewing subjects who unanimously described their movements in animate, causal, or intentional terms.[5] More recent studies have shown that subjects even arrange the space-time relationships between simple figures to reinforce their readings of intention—or simply as part of their inability to perceive events independent of intentional states.[6] And Alan Leslie's research conducted on six month-old babies has demonstrated that infants react with changes of facial expression, heart rate, and blood pressure when cinematic sequences representing noncausal relationships appeared sandwiched between sequences portraying causal connections.[7] Our tendency to perceive the world according to causal or intentional states seems endemic to our being human.

Human perception, however, also seems to work in the opposite direction, erasing "noise" when a strong signal is present, enabling us to hear only what we perceive as meaningful and to ignore anything extraneous that intrudes. Subjects in an experiment conducted by psychologist Richard Warren listened to spoken sentences recorded on tape on which a few phonemes had been erased and replaced by the sound of a dry cough. When asked to repeat what they had heard, the subjects overwhelmingly reported having "heard" the missing phonemes as precisely as if they had been present on the tape. Yet when an undoctored version of the sentences was played back to them, none of the subjects could distinguish the sounds they had imagined they had heard from the sounds they had actually heard— nor could they pinpoint the cough's location amid the sentences. As the cough existed outside any meaningful sequence for these listeners, their perceptions excluded it, but when a brief silence replaced the cough mimicking a speaker's pausing for emphasis, all the listeners were able to fix its exact location without difficulty.[8] The brain, apparently, does not preserve sequences step by step but, instead, recognizes overall patterns.[9] Our not necessarily conscious but prior recognition of these patterns enables us to unconsciously synthesize gaps such as missing phonemes without noticeable effort.[10]

It is our perceptions, then, and not the hand of a Scorsese or a Hitchcock that creates the illusion of continuity, sequence, and causation as we sit in a darkened cinema, showing us a steady sequence of moving images—where, in fact, only flickering, still images exist. As any cinema theorist will quickly inform you, nearly half the time we spend at the movies is, in fact, spent in the blackness between frames. Working together, the human eye and brain play off what Gestalt psychologists call the "phi phenomenon" and motion parallax to produce the fluid, tangible reproduction of life to which we have become accustomed onscreen.[11] What probably began as interpretive skills needed to ensure our survival in a highly competitive, natural environment now also endow objects and actions in the world around us with a continuity and a richness of meaning that enables aesthetic objects to exist and to retain meaning and significance for us.

Noting that Douglas Hofstadter has suggested the perception of relatedness, or connections, is perhaps the single, definitive characteristic of intelligent behavior, media theorist John Slatin argues that the link is the defining characteristic of hypertext:

> [E]verything in hypertext depends on linkage, upon connectivity between and among the various elements in the system. Linkage, in hypertext, plays a role corresponding to that of sequence in conventional text. . . . [T]he link *simulates* the connections in the mind of the author or reader.[12]

Hypertextual links or connections, of course, bridge the very physical gaps yawning between segments of text separated by virtual, three-dimensional space. Yet the links have no textual content themselves, and few cues that might prompt readers to see them as anything but a merely physical connection between two segments of text.[13] Although you might be able to see all the bridges between one segment and another in a hypertext narrative by peering at a map of it or browsing through menu options, no mutual set of conventions shared by authors and readers guides you in your choices. Some pathways might be given titles that are puns on their contents, some might even be helpful (as when the title to a path or place answers a question in the text), but their assistance to your navigation through the text is strictly a hit-or-miss proposition. In any case, attempts to paraphrase the contents of a fictitious passage are themselves problematic, hypertext or no hypertext, as anyone who has read Cleanth Brooks's "Heresy of Paraphrase" will tell you.

Yet if we recall the discoveries of the response theorists and psycholinguists of reading, we are confronted daily with gaps everywhere in the texts we read. They are even, critic Shlomith Rimmon-Kenan argues, inescapable:

> Holes or gaps are so central in narrative fiction because the materials the text provides for the reconstruction of the world (or a story) are insufficient for saturation. No matter how detailed the presentation is, further questions can always be asked; gaps always remain open.[14]

As Iser has succinctly noted, narratives represent opportunities for us "to bring into play our own faculty for establishing connections—for filling in gaps left by the text itself."[15] Gaps are precisely what enables us to engage in the act of "directed creation" so lauded by theorists from Sartre to Barthes, by leaving readers with what Sterne in *Tristram Shandy* called "something to imagine."[16] The reason why we so seldom glimpse these gaps—except in student writing, perhaps, or in our own writings in progress—is largely a function of human perception and only secondarily of literary convention. Conventions, as we shall see, are often little more than labels that we give to connections already, even inevitably, perceived between two objects or segments of text.

Text into Hypertext

Marc Saporta's *Composition No. 1*, a fictional narrative consisting of about 150 unnumbered, loose sheets of paper, is the first documented published work of fiction that required readers to compose the narrative themselves by shuffling the pages prior to reading them.[17] More radical in composition than Julio Cortázar's *Hopscotch*, which was to follow three years later, Saporta's text demanded that readers shuffle the pages of the narrative like a deck of Tarot cards, cut, and then read the cards to determine the fate of X, the protagonist. Similarly, in *Hopscotch*, Cortázar establishes two orders for reading his text, obliging readers to work out possible connections between chapters, such as location, chronology, and causality. In Saporta's narrative, however, readers are cast further adrift from the fixity of print, since the narrative, diced up into discrete cards, provides no such obvious script for readers to follow, leaving some commentators to conclude, as Bolter has, that the work never existed in a unified, complete form before its splintering into cards.[18] Looked at another way, Bolter's perception that the fragmented *Composition No. 1* may never initially have been a single, whole narrative owes less, perhaps, to what he sees as the elliptical character of the narrative than it does to the cuts introduced into the printed text.

Once you cut a text into fragments, essentially, you can liberate all the subversive, alternative connections possible between segments. To introduce physical cuts into a narrative and then present the whole thing to readers without any apparent order is to wrench the lid off Pandora's box and let rip the clamorous legion of voices identified by Barthes. At least, that was what I suspected when I sliced a short story into forty segments that ranged in length from a single word to three paragraphs. To test my hypothesis, I shuffled the fragments, stuffed them into envelopes, and handed these out to a class at New York University that consisted of twenty graduate students enrolled in a course in narratology. Organizing the students into ten pairs, I asked the class to place the fragments of the story in the order in which they thought it originally had existed, which enabled me to eavesdrop on the discussions, logic, and occasional squabbles between members of each pair as they worked their way, over an hour and a half of class time, toward a consensus about where each fragment "really" belonged.

As the pairs bent over their scraps of paper and began sorting through them, a surprising variety of strategies for assembling the text quickly revealed itself. Some teams began grouping similar actions they found in the text of the strips together ("There's an examination going on here and here: we have to put this one with that one"), and also clumping together fragments that had similar subjects or words in common. One pair placed three fragments together, all mentioning the names of cocktails or showing characters drinking. Most of the teams, however, began by reading each of the fragments and attempting to categorize the pieces according to literary conventions, with pieces sorted out according to

Their appearing to present "beginning," or "background" information

Their status as "kernels," elements essential to the plot, or "satellites," elements not essential to the plot, both structuralist concepts acquired during the course

Their presentation of "action" sequences

Their status as "exposition" or "descriptions" of characters or settings

Their appearing to supply "transitions" between sequences

The "voice" of the two characters in the story

The perceived "tone" of the narration

The sorting of narrative segments in nearly all pairs occupied the largest portion of the class time, and the pace of the assembly of the narrative, in most instances, quickened in direct proportion to the number of narrative fragments placed down in a narrative order agreed upon between the pairs. The more pieces a pair assembled, the more rapidly they proceeded to place the remaining fragments into a complete narrative order. Most strikingly, the teams appeared to rely upon similar strategies for constructing their respective narratives. First, they read through the individual fragments and attempted to articulate from them a global view of what the narrative might look like as a whole. Next, they attempted to find causal connections between actions or events from among the fragments to establish sequences or chronologies for what had happened. Finally, they "tested" these between themselves according to either their own life experiences or their knowledge of other narratives.

In the initial stages of assembling their narratives, nearly all pairs attempted to articulate between them some concept of what the work looked like as a whole, as what Teun van Dijk has called a "macrostructure," a global view of the narrative's themes and meaning.[19] For example, a number of pairs saw a fragment that read, "Two scenes ensue and we follow them both, you and I, *ma voyeuse charmante*," as a thesis statement, prompting them to organize the narrative into two scenes. Other teams perceived a pattern of alternation between the presentation of characters and ordered the narrative fragments strictly according to whether they featured the male protagonist or the female protagonist in them. And one pair decided that the story structure involved first a presentation of the male protagonist's problems, then the female protagonist's problems, followed by the interaction between them and, finally, exposition. These concepts of the narrative's macrostructure enabled the pairs to then establish scripts, or structures describing a sequence of events appropriate to a specific context—again, based upon their knowledge of human behavior and of literary conventions—enabling them to sort the fragments before them into some causal order and, after that, to establish narrative sequences.[20] As the pairs shuffled their fragments into sequences, they tried the order out on each other by telling narrative sequences aloud and applied their final, acid test to the narrative sequence or sequences after they told it. Each "telling" was subjected to the criterion of plausibility and/or a knowledge of narrative expectations based upon other works, their discussions showed:

"He wins her here, but she's annoyed at him here, so this scene has to take place after he wins her."

"If he operates on her and she gets mad at him because of it, then they wouldn't just be hanging out together like this."

"They get together at first, but then they fall out, so he has to win her back. That's how it ends."

After every group had completed their versions of the story, I provided each pair with a copy of the short story in its original form—and watched numerous sets of eyebrows rising and falling in disbelief at the difference between the scripts they had concocted and the scripts present in the original story. Several pairs complained how difficult it had been to read the story and to read it in different ways, in effect complaining that Bolter's requirement that we read in multiple ways forces us to "resist the temptation to close off possible courses of action; [and] . . . keep open multiple explanations for the same event or character."[21] The task, he goes on to note, is nearly impossible when we read print narratives because the medium itself, with its printed letters and words locked uniformly into lines, encourages us to think of it as a changeless, closed, and authoritative way of framing the events we see—as the *only* way of representing the events and characters we see within it (144). Even though Saporta's *Composition No. 1* is delivered to us in printed cards, piled in a box, our first instinct, ironically, is to find a way to put it back together again, to recover its "original" order. The static, fixed nature of the printed page and its austere linearity make it ideal for the presentation of a single order. Attempting to read a print story multiply—even one deliberately cut into fragments, or even, in the case of Saporta's narrative, overtly created as contradictory fragments—can never be more than an endeavor to fly in the face of the essential predispositions of the medium. Our perceptual bias toward seeing connections and causation enables us to create certain narrative sequences, as evidenced by the ease with which all pairs were able to arrange narrative fragments into causal clusters. The greatest difficulty facing the groups, however, lay in finding what seemed to be the definitive sequence that securely locked all the elements into place. If the print medium encourages us to see narrative events in a linear, singular, and definitive order, why did this last task prove so elusive?

First, *narratives themselves are filled with gaps.* As we have seen, gaps or indeterminacies in narratives are what enable readers to engage in transactions with texts, to breathe life into them. Since written language is itself indeterminate, there is no way for even the most "realistic" of narratives to represent characters and events completely concretely and wholly determinately. Many of the readers in my test group despaired of ever putting the story together, claiming that the original story itself must have been excessively, even incomprehensibly fragmented—only to express surprise at how coherent and even "obvious" the narrative sequence seemed, once they had read the original story. Good writing, as John Slatin has noted, involves presenting in print what seems to be not simply the most obvious, logical, and convincing order but, ultimately, the *only* possible order of events—one that seems "somehow inevitable."[22] Our tendency to perceive connections between objects and events in the text generally focuses on the way that the lines of print before us follow one another apparently seamlessly, locked into place. The physical substance of the page and the authority of the printed word combine to make these links seem definitive and to obscure any gaps yawning between actions or inferences. But when the text has been cut up, the gaps can threaten to engulf any glimmerings of coherence the narrative may originally have had.

Second, *all narratives have multiple connections, sequences, and orders other than the linear, syllogistic, or sequential order endemic to print narratives.* The very austerity of print with its singular order and relentlessly syllogistic, sequential connections, makes reading easier for us, in that it turns our predictions about where the text is heading and what to expect in the next sentence and paragraph into reasonably straightforward affairs. Other associative or thematic connections are driven beneath the surface of the text by its conventional linearity and tend not to haunt us until we are picking over the text very carefully, looking for little tufts of discord breaking through the surface. Although these alternative connections are available to us in retrospect when we read print narratives, we do not have to contend with them. Nor are we obliged to grapple, as we read, with the way in which these other, subversive orders may affect our perception of the narrative as a macrostructure. When these linear linkages are broken, however—as they were in the fragmented short story—readers are confronted with a plethora of possible and probable connections, boosted by our perceptual tendency to see links between characters and events that may, in fact, be entirely unrelated.

Third, *the fragmented print text prevented readers from "locking" into a single script or perceiving the narrative as a single, global macrostructure.* The "openness" of the fragmented narrative, with its many possible connections made overt by the cuts between paragraphs, made it difficult for readers to latch onto a specific script or schema that could enable them to assemble a singular and conclusive version of the narrative. Throughout the process, many were haunted by other possible connections, spurring them to invent explanations that rivaled the fiction itself as feats of imagination. Two groups insisted that the fragments were drawn from at least two different stories, making it impossible for them to create a single narrative order. One pair argued that the orphaned fragments they were unable to reconcile with the rest of "their" story had belonged to another story or stories and had somehow found their way into the envelope I had given them. And no fewer than four sets of readers claimed that crucial fragments had been left out of their envelopes, leaving them with gaps in their respective narratives that were unbridgeable. No readers, however, assumed that the fragments could have been constructed into multiple narratives that relied on the use of some segments more than once, since I had told them the bits of narrative in their envelopes came from a single print story. One group, perhaps taking their first, uncertain steps toward reading multiply, did, however, insist that their version of the narrative made richer, more probable, and satisfactory use of its contents than the original print short story.

For the readers overall, the "advantage" of the print narrative was that it held all the disparate connections and probable links between items, events, and characters in a single order that constrained ambiguity and supplied a limited and determinate closure for the events it described. But the print narrative also closed off other possible orders and alternate connections that clearly existed in the text, orders that were apparent when the narrative had been splintered into fragments. The pairs of readers could only attempt to lay the narrative fragments out in long chains of text (more than half the pairs actually spread the pieces out in strips on desks, tables, or the floor), because they were working with a print narrative, however fragmented. As we all know, print technology, as one of its chief "rules," does not allow you to include the same fragment in more than one narrative sequence, nor does it permit you to organize the fragments into mutually exclusive narrative sequences. Although many of the pairs were obviously reading multiply—as their complaints about having fragments from other stories in their envelopes attest—to read multiply in a print setting, to see an array of mutually exclusive possibilities for assembling the text, can only be disabling, a way of perceiving print text that does not lend itself to the discovery of the "right" narrative order. Nonetheless, nearly every one of the readers involved had, at some point or other during the exercise, managed to dredge up multiple possibilities for connections between blocks of text. In fact, many of the readers commented on the austerity of the connections in the uncut version of the story—probably because it represented a considerable reduction of flourishing links they themselves had perceived during the assembly of their respective versions of the story.

Two Kinds of Forking Paths

> In conventional narratives, readers are asked to imagine a
> world of multiplicity from within an overwhelmingly linear
> and exclusive medium. For hypertextual readers, the situa-
> tion is reversed—given a text that may contain almost any
> permutation of a given narrative situation, their task is to
> elicit a rational reduction of this field of possibilities that
> answers to their own engagement with the text.
> —Stuart Moulthrop, "Reading from the Map" (1991)

One might assume, perhaps, since hypertext has generated such a
buzz over its potential for reconfiguring the roles of author and reader,
that academia would be swamped under a tsunami of articles scrutinizing how readers handle hypertext. Instead, you would be hard
pressed to come up with an even dozen studies or considerations of
how hypertext may transform the way we read or write texts, and,
indeed, our whole conception of a satisfactory reading experience.
While you might find articles that examine the ways in which readers
can become disoriented in hypertext, mostly from the perspectives of
interface design and software engineering, you could also emerge from
a survey of the field wondering what happens when readers come face
to face with the technology. So, ironically, we have only the slenderest
knowledge of what is, quite possibly, the most important component
of the medium: the way in which readers interact with it. How do
readers conditioned to poring over linear narratives with tangible endings react when they run up against stories without endings, texts that
change with each reader and reading?

I thought "Forking Paths" represented as fair a litmus test as I was
likely to find—and not simply because it was one of only three hypertexts narratives in circulation back in 1987. Not only was the hypertext erected around a complex print story, providing common touchstones that could remain constant across readings of both the print
and hypertext narratives, but Borges's "The Garden of Forking
Paths"—as we already noted in chapter 3—was also a print story that
strains to escape from the confines of the medium. The class consisted
of freshmen writers in an honors expository writing course, held in a
networked Macintosh writing lab, the second of a two-course
sequence required of all New York University undergraduates. Without providing class members with any description of the experiment,
I divided them into two groups, with each instructed merely to read

either the print story "The Garden of Forking Paths" or the hypertext "Forking Paths" and then to recount their reading experiences and to retell the story they encountered in several paragraphs. For the first few minutes, half the class was originally disgruntled at being saddled with something as mundane as a print short story—many of the print readers ignoring the story to crane their necks in the hope of glimpsing what the others were doing at their Macintoshes. The other half was delighted at playing around with what looked like whizzy new software, until they ran up against a few navigational roadblocks. Admittedly, the software was, itself, problematic: this particular version of "Forking Paths" was accessible through an experimental, read-only module of a beta version of Storyspace, which barred readers from seeing either a map of the text or from browsing through lists of paths branching out from any particular place. Yet these factors alone are hardly huge obstacles that readers need to weave around: a number of hypertexts use a read-only form of the Storyspace interface similar to the "Forking Paths" version. What made blazing a trail through Moulthrop's hypertext difficult was this: "Forking Paths" did not contain any paths, strictly speaking, or defaults. Instead, it was constructed of a dense network of links that connected one segment of text to another through the conduit of words in the text. To continue reading, you needed to find words that would trigger links to other places containing related concepts, or the same word, or a similar word situated in a completely different context. If your choice of a likely candidate failed, however, to match the words included in the guardfield condition for the link—scripts attached to each link that required readers to select specific words from the text of any segment in order to move onward—the only action you would end up triggering was a frustrating beep from the Macintosh that informed you the word you selected was not attached to any link.

Unknown to either myself or my students at the time, Stuart Moulthrop was still in the process of writing full-blown instructions for readers of "Forking Paths," which he considered "crucial" to the reading of his interactive narrative. As he grappled with the hypertext, however, Moulthrop himself had forgotten that the read-only interface contained a set of control buttons enabling readers to move up, down, left, or right within the structure of the interactive narrative and to float around inside the hypertext independent of any connections between places.[23] To further complicate matters, "Forking Paths" had originally been created to run via a rudimentary text-linking program of Moulthrop's own devising that permitted only linked connections between nodes. When he decided to adapt the hypertext to a then still-provisional version of Storyspace, Moulthrop deliberately neglected to build default connections into the hypertext, intending his narrative to oblige his readers "to assume the mantle of authorship and expand the existing structure" (126).

Convinced that a hypertext structure such as "Forking Paths" would invite readers to become coauthors, Moulthrop had assumed that readers would become engaged by the text as structure at the level of words and phrases within each node, following the contentions of Russian Formalists and other theorists of reading—as formulated by Peter Brooks in *Reading for the Plot:*

> [W]e read the incidents of narration as "promises and annunciations" of final coherence, that metaphor that may be reached through the chain of metonymies: across the bulk of the as yet unread middle pages, the end calls to the beginning, transforms and enhances it.[24]

The most problematic aspect of this chain of metonymies, of course, is that readers—even those who most nearly match or realize the desires, intentions, quirks, and readerly history of an author's ideal audience—will not construe the same words as being meaningful, crucial, or intriguing in any given passage. It all goes back to the Iserian notion of the text as a quasi-determinate entity. When I read a word or phrase in the context of a compound-complex sentence, a sequence involving causation, or a novel with clearly defined characters and motifs, my concept of the connotative meaning of the word or phrase is sufficiently constrained to keep me from construing them in a far-out way. When, however, you ask me to relate my feelings about the word *war,* or *pink,* or *hatred,* the words tend to become something of a Rorschach test, prompting what can be a whole flood of associations, many of them totally unrelated to their immediate context in the narrative. As a result, initially, few of my students found themselves able to locate the words Moulthrop had selected as the anchors for his links—which meant they could not move through the text. Part of the reason behind Moulthrop's refusal to include defaults in his Storyspace structure stemmed from his desire for readers of "Forking Paths" to know when they had engaged a link. Had he incorporated defaults into the hypertext connections, readers would have moved through the text, regardless of whether they had activated a link or not. Certainly a "no-default" condition, as we saw in the last chapter, is considered integral to true interactivity, so that neither the author nor the reader of a hypertext has recourse to a single, predictable, and primary pathway through the hypertext network. Because of this no-default condition, however, what ideally should have promoted more interaction between reader and text stymied readers' efforts to engage the text on any terms but those the author had prescribed.

Predictably, the readers assigned "Forking Paths" became frustrated with what seemed to be odd software that just didn't work. Since they failed repeatedly to find words that yielded links, they complained that they could not read at all. Soon, however, one or two of the readers fiddling with the interface discovered that directional "up," "down," "left," and "right" buttons could pilot them through the virtual, three-dimensional space of the hypertext structure—completely outside the structure of links and projected interactions Moulthrop had scripted. The "right" directional button, quite helpfully, invariably yielded fresh places to read, and soon the readers were concluding that this was the only way they could move through the narrative at all, as if "Forking Paths" were a vast maze with dummy controls and lots of dead-ends. One reader noted that

> with very few exceptions, "right" was the only choice one could make in terms of movement within the story. The "up" option always took you back to the beginning, which was frustrating. . . . It was an interesting experience, and if there were more travel options (other than just "right"), I would have enjoyed it more.

Eventually, all seven of the readers given the interactive "Forking Paths" followed suit, using the "right" option to move through the text, and two ended up giving complete priority to the sense of the text they derived from navigating through the hypertext structure rather than their readings of the text contained in it. Confused by a multiplicity of narrative strands in which they could encounter a character dead in one place and very much alive and ambulatory in the next, the readers of "Forking Paths" drifted through the hypertext without any tangible sense of a macrostructure that could confer significance on the elements they encountered in any given narrative segment. Only by using their sense of the narrative as a virtual yet tangible structure could any of the readers arrive at a sense of the relationship between individual narrative places and their relation to the hypertext as a whole.

As hypertext theorists have discovered, perfectly acceptable hypotheses about what readers do with *Madame Bovary* are not a whole lot of help at describing what readers do with the likes of "Forking Paths." Moulthrop had constructed "Forking Paths" around Peter Brooks's conviction that readers entangle themselves in metonymic webs of language that ultimately lead them to metaphors for the text as a whole. But when Moulthrop sifted through my students' responses to his hypertext, he found them not reading for the plot so much as trying to plot their readings—struggling to establish where the particular places they read belonged inside the framework of the hypertext.

Rather than conferring a certain value on a scene because they perceived in it reflections of larger themes in the work, they instead attached significance to what they read because of the space it occupied in relation to the work as a whole. Neither of these reading strategies is, however, unique to reading novels or hypertext fiction. For example, let's say I am watching the last shot of the infamous shower scene in *Psycho*. As the camera spirals in toward Janet Leigh's lifeless, unseeing eye, I can begin to cobble together motifs in the scene (the cinematic equivalent of Brooks's metonymies)—the showerhead, the drain, the unblinking eye—perhaps recall others like it in earlier scenes, like Tony Perkins's eye fixed to the hole in the motel room wall, and begin to see the film as a cinematic treatise on voyeurism and madness, instead of a film about adultery and embezzlement, Hitchcock's famous red herrings that begin the whole enterprise. That is the metonymies-into-metaphor view.

I could, however, also decide that the film cannot just be about adultery or the wads of stolen bills still stuffed in Janet Leigh's suitcase when she is hacked to death only a third of the way through the film. Her murder is too explosive an event that occurs too early in the film for it to continue as a conventional thriller about adultery and stolen money. But her death is sufficiently early in the narrative for me to assume that the film will change into a story about the hunt for her killer or the other victims her murderer might rack up. We could dub this the "time-space" approach to reading. It is what enables us to chew our way unperturbed through a tub of popcorn as we watch Keanu Reeves hanging from the underside of a speeding bus in *Speed* because we know it is far too early in the narrative for anyone significant, let alone the hero, to die—unless, that is, he or she stars in a film directed by Hitchcock.

In any case, it is a strategy that is difficult to observe in your average print reader, since one always comes to a book with an absolute knowledge of how long it is, and a reader can, in any case, as Barthes reminds us in *The Pleasure of the Text*, always cheat a little, skip ahead, see what is coming next. What is striking, however, about this kind of interpretive maneuvering by the readers of "Forking Paths" is that they were guided by something they assumed but could not perceive. Although they could not see the map of "Forking Paths" while they were reading, they knew that its segments were like points on a map, as they could visit them by using directional tools. Since they had no clear idea of what the map looked like, their explorations were as much about getting a sense of the layout of the text and a map of narrative possibilities as they were about the placement and contents of any one segment within it. The map became, in a sense, a metaphor that represented the sum of the narrative's possibilities—turning on its head the print relationship between metonymy and metaphor suggested by Brooks.

Of course, print readers act somewhat like this—knowing that my edition of *Moby-Dick* has 536 pages in it does not help me understand what to make of Ahab's obsession with the whale or the significance of the doubloon the whalers nail to the *Pequod*'s mast, let alone Ishmael's relationship with Queequeg or the "Extracts (Supplied by a sub-sub librarian)" that preface the novel. And even knowing that this particular novel has been canonized as Great Literature and is about whales, or God, or the impenetrable face of Nature, does not make my reading of it as straightforward as swallowing ice cream. Just as we recognize what is happening in a scene from scripts we have learned elsewhere, we can use a script to guide us only when we can recognize its applicability to the scene at hand. Like readers of print fiction, their counterparts plowing through "Forking Paths" were simply doing what readers do: shuttling between microstructures and macrostructure, hopscotching back and forth between global structure, genre, and local meanings, as Jerome Bruner has observed:

> [A] reader goes from stones to arches to the significance of arches . . . goes back and forth between them attempting finally to construct a sense of the story, its form, its meaning. . . . As readers read, as they begin to construct a virtual text of their own, it is as if they were embarking on a journey without maps.[25]

Unlike print readers, who can draw on a vast repertoire of knowledge about genres, texts, and literary conventions, the readers of "Forking Paths" trudged through what seemed a trackless waste. What they could glean of the content—what actually happened in each segment of text—depended on an apparatus controlling their movements governed by rules that they could not quite fathom. And since they knew the hypertext was not linear, intuiting where the "ending" fell and what occurred in it would be pointless if it were not already impossible. One reader, however, made a game attempt to confer on the story some semblance of the orderly narrative strands and closure familiar from print stories:

> Yu Tsun later shot himself between the eyes but survived, he later became the ward of Herr Ignazius Baumgartner. Suddenly, he found himself in a garden maze, Madden and Stephen were also present in the maze, it is at this point, he had difficulty identifying his allegiance, was it to Elisa or Germany? . . . He leaves the room and found himself in the crossroad of the garden maze once again. Remembering dreaming about Amelie for the last nine nights, he confronted Madden and declared his renunciation of Berlin, the service and the chief. He was tired of the game, and so was Madden, [*sic*] Madden felt the same way. They became friends since they now share the same point of view.

As a synopsis of a conventional story, this seems reminiscent of the attempt by a contemporary of Sterne's who tried to take on *Tristram Shandy* by beginning with a gloss on the story and its plot and ended up wading through pages of description that seem destined, like Tristram's "autobiography," never to catch up to the task at hand. The summary of "Forking Paths" meanders along but seems to convey a basic sense of what the narrative is "about," which is certainly more than some of the other readers managed. This version, nevertheless, does violence to the narrative, skimming blithely across episodes that bring the narrative to (in some versions) a screeching halt—as when Yu Tsun shoots himself between the eyes. Had the reader continued along this particular branch of Moulthrop's web by alighting on the right words to trigger the next link, she might have discovered either that Yu's suicide attempt was entirely successful or that he had made a botch of it and ended up brain-dead (a possibility at least one other reader encountered). In her reading, Yu becomes a ward of Baumgartner—something possible only if the shot to the head has rendered him non compos mentis—reflects on his allegiances, bumbles through the labyrinth, opts to throw in the towel on the whole game of spying, and becomes pals with his former arch enemy, Richard Madden. Not bad going for someone who has taken a bullet squarely between the eyes.

What is remarkable about this reading of "Forking Paths" is the strenuous effort required to arrive at a synopsis that takes into account all its disparate and often mutually contradictory narrative strands and resolves them into a neat, linear series of events. To follow this script, though, this reader, probably abetted by years of reading knotty postmodern novels, has to brush away others, ones that insist that people shot between the eyes usually do not survive, much less possess the kind of powers of persuasion required to convince accomplished members of the opposition to also hang it all up. As might be expected, other readers displayed distinct signs of tetchiness with the narrative for straying from familiar print conventions. "This was a confusing story. People were killed, and then there they were talking in the next cell of the story," wrote one reader, while another complained: "I found dead people coming back to life." Neither of them, however, seemed to be able to glide over these episodes that followed each other spatially but related utterly contradictory events. Perhaps the reader who produced this relentless synopsis "saw" continuity and sequentiality because she had been conditioned to read that way almost from the afternoon when she stumbled her way through her first written sentence—like Richard Warren's subjects who "heard" the missing phonemes that had been replaced by the sound of a cough. Instead of seeing closure, or the end to a narrative strand, she read each ending as a transition, a bridge to another branch of the narrative, a little like a soap where this or that character's biting the dust merely clears the way for a change in narrative gear to follow tomorrow or next week. By stringing together a narrative of sorts, this reader manages to produce something like a compelling story of espionage, divided loyalties, and a resolution, a fair approximation of the satisfactions readers pursue in more conventional settings, where they "expect the satisfaction of closure and the receipt of a message," as Frank Kermode insists, by relying on the interpretive strategies we use daily in our face-to-face encounters: "To attend to what complies with the proprieties, and by one means or another to eliminate from consideration whatever does not, is a time-honoured and perfectly respectable way of reading novels."[26]

None of the other readers attempted anything like this Herculean effort to wrest "Forking Paths" into something resembling a linear story—although two were decidedly peeved at the contents of the narrative for not adhering to familiar print conventions. For them, the multiple endings and the dense, junglelike matrix of stories nullified any scripts they had for interpreting books, television shows, and films—but it did leave them with a script for working with something familiar: a maze.

Few of the readers had experienced a horticultural maze, where the ostensible objective of trotting around the topiary hedges is finding the center. Many of them had, though, stumbled across mazes in game books as children, where a confusion of paths separates a drawing of a treasure chest or a wedge of cheese at the center from a graphic of a pirate or a lean and hungry-looking mouse clinging to the outermost edge, and where tracing a wavering line through the maze to the center represents the solution. And, of course, the readers battling with "Forking Paths" at their Macs also knew that their classmates were turning the pages of a story entitled "The Garden of Forking Paths," which may well have suggested a garden maze, familiar from films like *The Draughtsman's Contract* or (more likely) *The Shining.*

In any case, two of them went on to work out a map for "Forking Paths" that would direct them toward the center of the fiction, where, presumably, all would be revealed—or enough to put them out of their misery. One of the "cartographers" struggled through his reading by trying to plot the placement of each chunk of text he trudged through, only to find that his concept of what should be where did not map all that brilliantly onto "Forking Paths." "When I did actually figure out where I was, where I had been, and where I was going, the story said otherwise," he wrote disgustedly. "In other words, whether I knew what path I was following or not, the story was too disjointed or random to comprehend." The other map plotter, however, through more advantageous choices, serendipity, or cheerful perseverance, apparently found a closer relationship between map, structure, and content—perhaps helped by his relying on a mental model of a maze, which provided him with a goal of sorts:

> At first, the stories did not seem to interconnect at all, but they all began to relate and make more sense the deeper into the story I got. After reading about 15 cells, I got to "<closure>," but I was not satisfied that it was over, so I continued reading. I made it to the center, cell 76, and that wasn't any big revelation, either, although I admit it was a little exciting at first. (I was wondering what I had won!)

Obviously, something that exists in virtual three dimensions has a shape, or a lack of something like linearity, for a distinct purpose: "Forking Paths" is, like haiku and Pound's *Cantos,* an intentional object, something constructed that signifies. Ergo, its shape must have some relationship to what happens in the narratives—making the center the best candidate for the Solution, something akin to the closing paragraphs of Faulkner's "A Rose for Emily," when the bad smells, the disappearance of her suitor, the immobility of the figure watching from the window are all resolved, all accounted for, all consumed by the surfacing of information we have been missing all along. Although he's an astute reader to have come this far, his script will inevitably fail him because it is based on, and abides by, laws governing the physical world. The maze can have only one center; stories (and lives) have only one ending. In one possible version of the text, the center actually does represent closure. In others, however, you can reach half a dozen endings without ever approaching it.

Looked at another way, however (since hypertext inevitably reminds us that there are always other ways of experiencing things), the reader may well have been learning a script as he read. The deeper he penetrates into the web of the hypertext as he shuffles to the right, the more the segments seem to be resolving into fragments of sequences, albeit splinters of several different narratives encountered in a mostly higgledy-piggledy fashion. He recognizes connections but cannot arrive at resolutions to the narratives that swim past him, so he tries to look to the external structure of the text for clues. When these prove elusive, however, he resorts to his own definition of closure: visiting most of the points on the map he's constructed of the structure of "Forking Paths." His decision to declare his discovery of closure on his own terms is worth wondering at, since it has no precedent. While it might be kosher for me to decide that the ending of *Jacob's Ladder* is completely unsatisfactory and nullifies rather than accounts for what preceded it, it would be well-nigh impossible for me to declare that I had found a more satisfactory version of closure in the film.

Closure in stories, novels, films, and television series, even when it is left open to future episodes or sequels, is always determined by authors, screenwriters, directors, and producers. It is not something we as readers can take or leave. But, if readers realize they are dealing with possibilities and versions, rather than events that are immutable and determined, they also need to account for why they finished their readings. With the "Forking Paths" exercise, several of the readers confessed that they stopped when they became too frustrated or confused; some claimed they had simply run out of time; and a few argued that they had reached a sense of closure. When you reach the end of *Middlemarch* or *The 400 Blows*, although you can dicker over whether the ending was satisfactory or a disappointment, you cannot define closure as an abstract entity: it is an integral part of any published poem or piece of fiction. When you reach the "end" of "Forking

Paths," or *afternoon* or any other piece of hypertext fiction, however, closure becomes an entity that needs a lot of defining.

Endings, Closure, and Satisfaction

> We might say that we are able to read present moments—in literature and, by extension, in life—as endowed with narrative meaning only because we read them in anticipation of the structuring power of those endings that will retrospectively give them the order and significance of the plot.
> —Peter Brooks, *Reading for the Plot* (1985)

"The sense of adventure . . . plotted from its end," Peter Brooks writes, is part of a pattern of "anticipation and completion which overcodes mere succession."[27] Put another way, our expectation that *Presumed Innocent* will end with the revelation of the murderer's identity—as nearly all thrillers and mysteries that include corpses between their covers do—controls how we read the story from its beginning onward. Did he, or didn't he? we wonder. Could she or couldn't she?

Figuring out how something might end, the direction in which the narrative seems to be barreling along, helps orient us at the beginning, particularly when the signposts along the way are far from clear, as they were in "Forking Paths." Like the reader who fixed onto the concept of the text as a literal labyrinth or maze and headed straight for its center, our forming hypotheses about the kind of knot that will tie up the textual loose ends helps us interpret most of what happens along the way. Endings, or knowing how to find them, provide us with a goal, enabling us to sift through episodes and separate the distinctly goal-related ones ("He definitely did it: he was jealous of her sleeping around, particularly with his boss") from those that seem extraneous to the outcome of the plot ("So his wife is bright and frustrated—who cares?"). What seems, perhaps, most peculiar about the "Forking Paths" exercise are the reactions of the undergraduates who spent the class period poring over the Borges print story. They not only felt the ending of the print story was overdetermined, two readers even argued with its viability, insisting their own versions fit the outlines of the story's scheme far better:

> I figured out Tsun's [*sic*] motivation throughout this story after rereading the beginning of it, but that does not seem like

the point that [the story] was really trying to get put across. For all we know, the particular end used in this story may have been only one of an infinite number of possibilities down the "forking paths" of time.

Since, according to the all-wise Ts'ui Pên, all of these possibilities peacefully co-exist in parallel dimensions, isn't it true that we can reject this ending and make up one of our own? Dr. Stephen Albert even tells Tsun, after receiving his reverence, that in one possible future, the two men are enemies, and Tsun lamely replies that "The future already exists," lies to him, and then kills him; and all just to send a message to his German "Chief" whom he hates to prove the point that a "yellow man" was capable of saving the German forces.

This reader has decided that the spy plot, which provides the ostensible raison d'être for the story, is not really the subject of "The Garden of Forking Paths." The real story is about labyrinthine possibilities that exist in time, he argues, pointing out that Yu lacks sufficient motivation to shoot Albert dead—or, at any rate, that this is one of the less satisfying outcomes possible in the story. Since the story is all about labyrinths like Ts'ui Pên's, where "all possible solutions occur, each one being the point of departure for other bifurcations,"[28] it seems illogical for the story to end abruptly with Yu killing Albert and getting hanged himself for the crime.

The other resistant reader, having decided that the whole plot is riddled with inconsistencies (of which the ending is evidently the most glaring and frustrating example), homes in on what seem like logical flaws in the short story:

> Tsun [*sic*] then kills Albert, and is captured by Madden to be sentenced to death. But in killing Albert he is able to convey the name of the city that the Germans have to attack to survive, which they did (attack, that is). But some questions remain: Why didn't Tsun just kill Albert and then leave, instead of remaining with Albert long enough to allow Madden to catch up with him? Why did Tsun only carry one bullet in his gun? Why did Tsun declare to Albert, just before he killed him, that "I am your friend"? . . . In the definition of the book, is Albert still alive, just living one of his other lives?

A careful and insightful reader, after reading the story through to the ending, she goes back again to the beginning, a fragment that puzzled her the first time around: a paragraph that paraphrases *A History of the World War* about the postponement of a planned offensive that seemingly lacked any special significance—a pause that can be accounted for by Yu Tsun's deposition. This entry "takes us nowhere (apparently)," she writes. "After reading the story and a little thought, I decided that maybe the author was implying that the outcome of the battle was inevitable." The Germans survive only to engage in the battle described in the fragment that begins the story because they manage to bomb the British artillery park, and they know the location of the park only once Yu murders Albert and the story makes news headlines. The Germans, however, can connect Albert's name with a valued piece of intelligence—the location of the bomb depot— only if they know that one of their spies has killed him, so Yu must allow himself to be captured—a logical connection this reader cannot quite make, possibly because she sees the story as violating its own precepts. Ts'ui Pên's book, "The Garden of Forking Paths," after all, insists that time exists in an infinite series that embraces "every possibility," as Stephen Albert tells Yu, and Yu agrees with him—even senses phantom versions of the two of them "secretive, busy and multiform in other dimensions of time."[29] So doesn't this mean, this reader demands, that Albert is just living one of his other lives? The relentlessly conclusive spy story is, after all, only one of the many possibilities that exist within the garden of forking paths, the labyrinth created by Ts'ui Pên and rediscovered by both Stephen Albert and Yu Tsun. In the Borges narrative, the garden enables mutually exclusive possibilities to exist simultaneously, and it is the garden that becomes the metaphor for the "Garden of Forking Paths," not Yu's murder of Albert or his own subsequent execution. It seems that the metaphor of the garden, with its plethora of possibilities, invites readers to decide that their own endings to "The Garden of Forking Paths" can assume importance equal to the ending conferred on the narrative by Borges. In this instance, the print readers climbed through the text via a chain of metonymies, exactly as Brooks described, guided by a metaphor that represented the text as a totality—an outcome that makes one realize that, for all the aggravation reading "Forking Paths" caused these readers, Moulthrop was on to a good thing after all.

In the end, we are left with something of a paradox. The readers of the interactive "Forking Paths" seem swamped by a multiplicity of endings and narrative possibilities. The readers of the print story "The Garden of Forking Paths" appear uneasy with Borges's singular and very limited ending to a narrative that concerns itself, nearly to its penultimate paragraph, with a universe of seemingly infinite possibilities. We can explain the differences in their feelings, perhaps, by taking into account the differences between the media. After all, the hypertext readers were confronted by a text that seemed to blithely transgress every convention, every expectation familiar to them from their excursions in print. On the other hand, the print readers were merely reading a rather problematic print story, hardly a novel experience to undergraduates taking English courses. Yet the group of hypertext readers included one who succeeded in charting a map of the structure of "Forking Paths" and used it to confer meaning upon narrative segments according to their placement within its structural space—hardly the response of a reader overwhelmed by the demands of a new and entirely foreign reading environment.

In fact, the reader of "Forking Paths" who stressed navigation, together with the readers resisting the ending supplied in "The Garden of Forking Paths," seem to have arrived at a definition of closure that revises the usual traditional definitions, where it is generally viewed as both the place where "goals are satisfied, and the protagonist [can] engage in no further action" and the "point at which, without residual expectation, [readers] can experience the structure of the work as, at once, both dynamic and whole"[30] Whereas traditional definitions of closure generally give priority to our grasp of the structure of the work as a whole, these readers seem to have arrived at strong readings that emphasize their sense of the text as part of an ongoing dynamic—a quality that Nancy Kaplan and Stuart Moulthrop first identified in a study of readers of interactive narratives. As one of the readers in their study "Something to Imagine" noted, closure occurs "when we have decided for ourselves that we can put down the story and be content with our interpretation of it. When we feel satisfied that we have gotten enough from the story, then we are complete."[31]

Similarly, all three "strong" readings of "Forking Paths" or "The Garden of Forking Paths" featured what we might call a "reader-centered" dynamic. The reader navigating through the narrative spaces of "Forking Paths" decides, based on his sense of having traversed both the margin and center of the hypertext, that the text will not hand him any great resolutions that might resolve the tensions and possibilities percolating in its plethora of narrative strands. Reflecting back on the dynamic universe of possibilities set up by Ts'ui Pên's Garden of Forking Paths, the readers of "The Garden of Forking Paths" feel that the narrative continues unfolding beyond the singular ending outlined by Borges. "In the definition of the book, is Albert still alive, just living

one of his other lives?" wonders one reader. The other reader, however, perhaps realizing that classmates engaged with the electronic
text were experiencing alternate endings to the same narrative episode
he encountered, does not simply resist the singular closure of the print
story—he invents his own ending:

> I prefer a future in which Tsun [*sic*] murders Richard Madden
> and lives peaceably with his mentor Dr. Stephen Albert, fol-
> lowing in the footsteps of his ancestor Ts'ui Pên and finishing
> his life work: "The Garden of Forking Paths."

Here it is not simply a case of dealing with two different media
environments, since resistant and reader-centered interpretations
alike emerged from readings of both the print and hypertext narratives. We appear, instead, to be dealing with two different kinds of
readers: what David Riesman, in *The Lonely Crowd,* once dubbed
"inner-directed" and "other-directed."[32] Inner-directed readers here
are distinguished by their ability to redefine their roles as readers
either through discovering a new way of navigating through narrative space or by revising the concept of closure. Other-directed readers, conversely, take their cues for reading from their knowledge of
established reading practices and literary conventions, leading them
to brand examples of narratives wildly divergent from familiar
norms, such as "Forking Paths," as frustrating, nonsensical, and,
even, failed.

This distinction between the reactions of inner- and other-
directed readers, of course, hardly applies simply to hypertext, just as
it does not relegate hypertext fiction to the realms of distinctly high-
brow, avant-garde works. Just as we do not expect fiction or novels to
belong to some monumental genre, hypertext or hypermedia fiction
will likely expand to fit the demands of their audiences: print fiction
is not all *The Bridges of Madison County* any more than it is all
Finnegans Wake. Readers who prefer the neat firmness of endings can
already find the pleasures of anticipation and confirmation ticking
away in digital narratives like *The Magic Death*—albeit with the nagging realization that, no matter how satisfied they are with an ending,
it could always be (and will probably, on further readings, turn out to
be) otherwise. No matter how you look at it, closure becomes a handful the moment you remove it from the category of strictly necessary
things we cannot choose to do without, like death and digestion.
Before hypertext came on the scene, it was a commonly acknowledged

fact that everything had to end, making endings things that were either satisfactory or unsatisfactory but not, in any instance, debatable. Once we dispense with closure as an entity that is always determined by an author and always consumed by a reader, however, we can clear for ourselves a bit of neutral ground to examine what defines an ending beyond the blank space accompanying it.

Just Tell Me When to Stop: Hypertext and the Displacement of Closure

[Conventional novelistic] solutions are legitimate inasmuch as they satisfy the desire for finality, for which our hearts yearn, with a longing greater than the longing for the loaves and the fishes of this earth. Perhaps the only true desire of mankind, coming thus to light in its hours of leisure, is to be set at rest.

—Joseph Conrad, "Henry James" (1905)

Death is the sanction of everything the storyteller can tell. He has borrowed his authority from death. In other words, it is natural history to which his stories refer back.

—Walter Benjamin, "The Storyteller"

Just how essential is closure to our readings of narratives? Do we read narratives to satisfy some deep craving for closure denied to us in our everyday lives, as Conrad and Benjamin argued, an aesthetic equivalent that entitles us, godlike, to have the whole story revealed to us nakedly, in its entirety, something we seldom see in life? Part of the pleasure in reading is tied up with an ending—with our knowing that, no matter how many characters wander through the pages of *War and Peace*, it is strictly finite: it all ends, satisfying some longing or curiosity within us and freeing us to pursue other things. Or perhaps closure is more than simply a mere component in our pleasure, perhaps it is integral to narrative aesthetics and poetics, as critics like Peter Brooks and Frank Kermode have persuasively argued.

Using the sentence as a paradigm of narrative structure, Brooks argues that in narratives "the revelation of meaning . . . occurs when the narrative sentence reaches full predication."[1] Narratives without closure are like sentences that include only the subject and not the "action" of a sentence, the predicate. Closure, in this view, does more than complete the meaning of a story. It limits it, much in the way that my using the word *plink* as a verb in the sentence "The note plinked," tells us that the note is musical and not, say, a billet-doux. Instead of wondering whether the noun "note" means either a paper mash note, or a piece of paper currency, or a musical tone, I simply infer that it means a musical note. As Brooks and psycholinguists of reading would have it, my inference drives me forward, looking to confirm my unconscious hypothesis; the predicate "plinked" restricts my possible readings of "note" to sounds made by musical instruments—and tells me I was right all along.

Anticipation helps us comprehend ambiguous and indeterminate language, which includes just about everything in the realm we classify as fiction or literature, not to mention the gamut from the *National Enquirer* to sales stickers in Macy's. But it also provides us with boundaries that confine our predictions and interpretations: I am scarcely aware, if at all, of pausing between reading "note" and "plinked" to mull over the possibilities. "Only the end can finally determine meaning, close the sentence as a signifying totality," Brooks claims, insisting that our predictions or projections of closure, like the predicate in a sentence, enable us to interpret narratives.[2] But we cannot be quite certain that our hypotheses are right, and, in fact, we may simply wonder what everyone's trying to warn the Stephen Rea character about in *The Crying Game,* and not formulate any theories at all about the gender of the object of his infatuation. We are merely aware, at least momentarily, of the narrative's being left deliberately, teasingly incomplete. Reading is, more than anything, an act of faith, belief that, by the time we reach the ending, everything we have witnessed will at last make perfect sense, all our nagging questions will be answered, all disputes settled, all the wayward threads corralled into a tidy unity: something we can lay to rest before comfortably turning our backs on it. When endings are looked at in this light, it is not difficult to account for our seemingly inexhaustible desire to reach them in the stories we watch or read.

While the inner-directed readers of "Forking Paths" and "The Garden of Forking Paths" alike generated their own sense of closure in the absence of an acceptable ending, the fact does not tell us much about how the suspension of closure affects our ability to understand narratives at a local level. After all, if I am not certain where my destination lies or how I will recognize it once I reach it, how am I supposed to distinguish the landmarks that should guide me into the narrative equivalent of a safe harbor from the detritus along the wayside I should simply ignore? Questions also remain about how integral closure is, not only to our ability to make sense of narratives, but to the pleasure we take in reading or watching them. Do we read for something that resembles closure, even when endings themselves are an impossibility? Or do we perhaps simply redefine closure as the reader of "Forking Paths" did? Or do we just create our own list of the probabilities we find most plausible or palatable or logical, as the readers of the Borges short story did, and call it closure?

Classical Closure and Twentieth-Century Print Narratives

> Really, universally, relations stop nowhere, and the exquisite problem of the artist is eternally but to draw, by a geometry of his own, the circle within which they shall happily *appear* to do so.
> —Henry James, preface to *Roderick Hudson* (1874)

> I have, I am aware, told this story in a very rambling way so that it may be difficult for anyone to find their path through what may be a sort of maze. . . . [W]hen one discusses an affair—a long sad affair—one goes back, one goes forward. . . . I console myself with thinking that this is a real story and that, after all, real stories are probably best told in the way a person telling a story would tell them. They will then seem most real.
> —Ford Madox Ford, *The Good Soldier* (1915)

It is no coincidence that critics such as Peter Brooks, Frank Kermode, and Walter Benjamin insist on closure as an essential component—perhaps *the* essential component—in narrative poetics. Contemporary concepts about the role of endings or closure derive authority from Aristotle's simple definition of story as having a beginning, middle, and ending. For Aristotle, the definition of plot, or what we might call "story," is "a whole . . . [with] a beginning, a middle, and an end," where the beginning "does not itself follow anything by causal necessity," and the ending "itself naturally follows some other thing, either by necessity, or as a rule, but has nothing following it."[3] In the same vein, Kermode argues that the provision of an ending "make[s] possible a satisfying consonance with the origins and with the middle," thereby giving "meaning to lives and to poems."[4] But, Kermode goes on to insist, the ending need not be provided by the text itself (or announced by a lengthy newspaper obituary) in order to endow meaning on the life or narrative that has proceeded it because, as readers of texts and of lives, we create "our own sense of an ending" by making "considerable imaginative investments in coherent patterns."⁵ Endings, in other words, either confirm or invalidate the predictions we have made about resolutions to conflicts and probable outcomes as we read stories, watch films, or speculate about the secret lives of the couple across the street.

While Kermode's "coherent patterns" dimly echo Brooks's flow of metonymies, they also suggest theories of reading that enshrine readerly predictions about what word or event is coming next as the single action that enables comprehension. I cannot read a sentence without making inferences: about what a pronoun refers to, which denotative meaning is intended with words that can signify radically different things, what certain words connote, how this line of text fits in with the others I have read, where the writer seems to be steering the prose. These hypotheses help me limit ambiguity at a local level and, some psycholinguists of reading would insist, at a global level as well.⁶ As Kermode's theory of interpretation would have it, however, we are not merely looking perpetually forward—we are forever envisioning the end. Like Brooks's "anticipation of retrospection,"⁷ Kermode's act of reading is endlessly recursive, continually building a structure that presupposes an ending that, in turn, modifies the building of the structure. Brooks takes this still further, making closure the limitation on narrative that defines its shape and significance:

> [A]ny narrative plot, in the sense of a significant organization of the life story, necessarily espouses in some form the problematic of the talisman: the realization of the desire for narrative encounters the limits of narrative, that is, the fact that one can tell a life only in terms of its limits or margins. The telling is always *in terms* of the impending end.⁸

Significantly, Brooks, Kermode, and Benjamin use closure as the single entity that confers cohesion and significance on narratives in a way that strongly suggests that the experience of narrative closure numbers among the principle pleasures of reading narratives—the one thing that both prompts and enables us to read.

It is, perhaps, no coincidence that all three of these critics also typically concern themselves with what we might define as "classical" narratives, texts that predate the modern and postmodern eras. Although Kermode touches briefly on Robbe-Grillet, acknowledging that the "reader [of Robbe-Grillet] is not offered easy satisfactions, but a challenge to creative co-operation,"⁹ he concerns himself chiefly with fictions that have determinate closure—endings that are paradigms of an apocalyptic and definitive end. Discussing Robbe-Grillet's *In the Labyrinth,* he is only with difficulty able to grapple with its representing a conceptual labyrinth that continually violates our expectations of narratives—a text that provides none of the continuity, coherent patterns, or closure endemic to works from which (and upon which) Kermode bases his textual aesthetics. "[T]here is no temporality, no successiveness. . . . This is certainly a shrewd blow at paradigmatic expectations," Kermode writes, then dismisses the Robbe-Grillet's work as simply "very modern and therefore very extreme" (21).

As any sociologist will tell you, it is far easier to predict practices represented by common and highly conventionalized examples than it is to guess how the woman or man in the street might make sense of an object that aims to deliberately subvert them. It is one thing to theorize how readers behave in the throes of reading *Joseph Andrews* or *Nana* and quite another to account for what any of us might do when confronted with the likes of Coover's "The Babysitter," or Cortázar's *Hopscotch,* or Fowles's *The French Lieutenant's Woman*—all of which contain multiple and therefore highly indeterminate endings. In the eighteenth- and nineteenth-century novels and stories most critics use as fodder for their theories, plot and narrative are generally bound inextricably together. But in twentieth-century fiction, stories may "end" long before the narrative finishes on the last page of a book, making it difficult for us to perceive the "ending" to which Brooks or Kermode would have us refer. Which ending to *The French Lieutenant's Woman* should I be anticipating: the "traditional" paternalistic resolution, or the less resolutely cheerful, more "modern" one? And can I make inferences or draw conclusions comfortably or with any degree of certainty, if I know that there are at least two endings lurking toward the back of the book? Far from strictly limiting ambiguity, multiple endings would seem to leave things open—certainly far more than conventional endings.

While it is certainly true that readers of Jane Austen's *Northanger Abbey* proceed through the novel wondering if Catherine will ever be united with her beloved Henry (and probably view nearly every scene in the book according to this light), I can work my way forward through Ford Madox Ford's *The Good Soldier* already knowing, perhaps not the end of the narrative itself, but certainly the "end" of the story, the events that take place at the very limits of its chronology. I do not need to anticipate the end of the plot: I already know it. The narrator, John Dowell, has told it to me, a mere two-thirds of the way through the novel. Both Edward Ashburnham and the narrator's wife, Florence, are already dead, and Dowell's beloved Nancy, mad and vacant, has been entrusted to his care. Since a series of flashbacks shape Ford's novel, the narrative rockets between two separate chronologies, one that traces events in the friendships and marriages of the Ashburnhams and Dowells and one that follows Dowell's slowly burgeoning awareness of what happened in the interstices of the first chronology. On one level, *The Good Soldier* is about the old themes of adultery and deceit; on another, it is the chronicle of a loss of innocence or naïveté, the destruction of Dowell's utter faith in the world of appearances. So closure is not really present at the "ending" of the plot that, in any case, grinds to a halt before the last third of the novel. Closure here is something the narrative provides as Leonora's bitterly perceptive memories eventually penetrate Dowell's blissful ignorance, transforming his memories and history alike. If I had not initially suspected Dowell was a naive observer who, as a judge of the events he perceives, falls somewhere between Benjy in *The Sound and the Fury* and the child in Henry James's *What Maisie Knew*, by the time a bystander remarks on Florence's not-so-salubrious past in Dowell's hearing, I am already a good fifty pages ahead of him. What keeps me plowing on is my curiosity to discover just how far the wandering eyes of Edward and Florence will stray, and exactly how long it is going to take Dowell to piece it all together. I already know how it all ends—I am just waiting for Dowell to catch up, as he does, a measly eight pages from the end of the book.

Similarly, what are readers to make of Robbe-Grillet's *In the Labyrinth*, which continually reverses our expectations from sequence to sequence, and from paragraph to paragraph—and even, occasionally, from sentence to sentence? A soldier walks through the streets of an unnamed town, carrying a box. He is lost; he is in his barracks dormitory. He is merely tired; he is mortally wounded. He is a figure in a photograph; he is a figure in an engraving; he is a soldier trudging through snowy streets. The engravings and photographs come to life; the sequences we read may or may not have happened—in fact they may not even be probable. At the end of the narrative, a doctor identifies the contents of the dead soldier's box; at the end of the narrative, the soldier and his box appear in an engraving and the narrative takes up again where it first began, with descriptions of the interiors of dusty rooms, and snow falling silently outside. In Robbe-Grillet's novel, as well as in *The Good Soldier*, closure in the conventional sense has been displaced. The novel's end, like a labyrinth, simply draws us back to its beginning without either confirming, negating, or resolving the tensions, questions, and hypotheses we bring to our reading of the narrative. Whatever the narrative offers in the way of goal seeking—the soldier's attempt to orient himself in a strange location, or the mission behind the box he clutches to him—is never resolved (or even addressed directly) in the narrative. Even referents for pronouns may change within the space of a paragraph, making it well-nigh impossible for readers to guess about future events or to establish a sense of the causal relationships between characters' actions or narrative episodes—in short, to perform any of the actions readers normally do.

When these relationships are violated at every turn, as they are in *In the Labyrinth*, we can call upon our knowledge of narrative conventions to hold our reading of the text together. As readers we expect characters to remain constant throughout the narrative. We do not, for example, assume that the soldier we follow through the streets will metamorphose into someone else as we follow him (unless we know we are reading horror or science fiction)—as he does in Robbe-Grillet's novel. We are accustomed to shifts in time and place being signaled by transitions or descriptions that pursue characters as they move from one setting to another. And we wait to learn about the most important events in the story as we meander through the narrative. But in Robbe-Grillet's narrative, all these expectations, all this readerly patience seems profoundly misplaced. I discover the soldier is wounded without having learned just how or when, and, in the end, I am not sure that he existed at all, leaving me perhaps less certain about the status of characters and events than I was even at the outset. The ending of the novel prompts me to recognize its structure as a textual labyrinth (something I am becoming increasingly familiar with), but it is the continual subversion of my expectations that gradually induces me to see the narrative as a form of antinarrative, a gesture that reveals to me the nature of the unseen elements for which I unconsciously search as I read—without delivering to me the actions, consequences, or resolutions I overtly seek. The ending, to use Barbara Herrnstein-Smith's definition of closure, simply removes any "residual expectations" I may have—I know that the narrative has nothing left to reveal after I have finished my reading of it and that I am free to begin to make sense of the work as a whole.[10]

So, to echo Gertrude Stein, an ending is an ending is an ending. Regardless of whether the plot stops dead or dribbles to a halt in the middle of a novel, or the narrative turns itself into a labyrinth where neither the center nor the end offers us any scraps that resolve riddles in the plot, their physical endings sever our connection with the story.

We can only reread them, either now, haunted by the same unanswered questions and nagging doubts, or later, as a slightly different reader who has either forgotten all the old questions or has evolved entirely new ones. In any case, the text, as Plato noted so aptly, will just keep saying the same thing. Nevertheless, simply knowing that an ending exists, an all-will-be-revealed implicit compact between the reader and absent writer, invites us to make projections, assures us that our inferences will be held up against the real article before the pages run out. But what happens to me when I plunge into *afternoon* with its thicket of segments, all 539 of them, and its 905 connections—let alone *Victory Garden*, with nearly a thousand segments that I may read or possibly never get to, strung together with twenty-eight hundred defaults, links, and paths? Even a digital narrative like *The Last Express*, which, as a "story-based adventure game,"[11] offers only one "winning" ending, also offers eleven "losing" endings—some of them more neatly resolved than the "winning" ending—as well as narrative threads, overheard lunchtime conversations, romance, and backstory on its cast of characters that in no way determine the outcome of the game-plot. In fact, readers tempted to eavesdrop on conversation between two characters lingering over a late lunch will miss an opportunity to ransack their compartment for clues that may or may not help resolve one of the plot's central puzzles. Readers can successfully negotiate the plot's challenges and arrive at the "winning" conclusion without experiencing half the content of *The Last Express*[12] even after twenty-five hours or more with the narrative, a rare experience in our age of obsolescence—and the sort of satisfaction even readers of *In the Labyrinth* can enjoy.

How can I read without fixing my sights on an ending—any ending? Whereas readers of even the most "difficult" writers in print face texts already supplied with endings, readers of hypertext fiction generally must supply their own senses of endings. Since the mechanisms necessary for me to interpret a novel resemble those that enable me to read a sentence, an exploration of how hypertext readers deal with the suspension of closure might cast light on the relationship between structures integral to the act of reading and the concept of closure. What prompts readers to decide they are "finished" with a particular interactive narrative and to discontinue their readings of it? And can readings, a cumulative experience of the narrative's manifold possibilities, approximate a sense of closure for readers, a sense they have grasped the narrative as, in Jay Bolter's phrase, "a structure of possible structures," even though their readings may not have explored every narrative space and connection?[13]

Looking for the Close of afternoon: *Four Readings*

> I want to say that I may have seen my son die this morning.
> —Opening segment of *afternoon* (1990)

Hypertexts, of course, do not have pages, and, in any case, a mere tally of how many places *afternoon* or *Victory Garden* contains tells us little about how long any one reader might spend with it, since you may encounter the same two or five or fifty segments in more than one context—or run through the entire hypertext without twice stumbling across the same segment. In any case, the length of time you or I might spend reading anything in print is a poor measure of the time and effort involved in exploring a hypertext, especially since readers of hypertext narratives can spend up to six times the length of time required to read print narratives.[14] So a single reading of *afternoon* could occupy the same amount of time as a reading of an entire novel such as *The Good Soldier*—or, in some cases, what can even amount to entire months of reading, the kind of time you might lavish on Proust or the Bible. Conversely, depending on the paths you take through any hypertext, one reading can correspond to the time you might spend with a single chapter of *Lord Jim*. With no clear-cut divisions such as chapters between episodes or narratives strands, readers of interactive narratives encounter few cues telling them where to pause in their reading for a breather, let alone when they have completed one possible version among the narrative's multiplicity of stories.

This does not mean that hypertext narratives are inevitably, like Antarctica, without the usual signs or paths we need to guide us. There are, after all, certain limitations authors can build into their hypertexts to encourage behavior, foster conclusions, or invite assumptions from their readers—as we saw with Moulthrop's building links from words into the design of "Forking Paths." The simplest form of these also most closely approximates the limits familiar to us from print: by removing default options from the connections or by attaching a condition to its paths, the author makes it impossible for readers to move beyond a given place. So if I were working my way quickly through a hypertext for the first time solely by relying on defaults, the hypertext equivalent of channel surfing, stumbling into a place that has no default connection from it can seem tantamount to running headlong into a brick wall. Or the ending of the narrative. It is entirely up to you, dear reader.

Afternoon has a network of connections sufficiently rich that, even if I were to read several hundred places strictly by default for two readings, I could produce an entirely different version of the narrative simply by choosing a path instead of a default—say, by answering "yes" or "no" when the text in "Begin" queries, "Do you want to hear about it?"—in only one of its places. In certain lights, reading by default can produce the most accessible reading of *afternoon,* since I am simply pursuing an already determined path, one with an invisible logic and no apparent branches off it that does not require me to mull over which paths or links I should be looking for, just like a book. Fittingly, this version is also literally the most accessible, hewing to the contours of plots about quests. The narrator, Peter, fears that he may have seen the bodies of his estranged wife and son lying by the roadside as he drives into work. He drove by too quickly to get a fix on their identities, so he might be wrong, but when he embarks on a series of phone calls to reassure himself that both are still ambulatory and going about their everyday business, he uncovers nothing but gaping holes: no one has seen them. Panicking, he scouts around the accident scene—and uncovers a scrap of paper, a school report in childish writing he recognizes as belonging to his son, blowing around on the grass. The pace of his now-frenetic quest heats up, then ends abruptly thirty-six places into the hypertext, with the narrator deciding not to begin phoning around the local hospitals but, instead, to call someone named Lolly. No default branches out from this place, and, moreover, not even the menu of paths is accessible from this point, at least not during this particular reading. Since Lolly has not yet appeared in the narrative, I have no idea what significance she holds for Peter's quest. The possibility, also, that she might hold the key to the whereabouts of wife Lisa and son Andrew—since the narrator's last resort before calling around the local casualty wards is to take a tranquilizer and call her—makes this version of the narrative seem particularly inconclusive.

During my second reading of *afternoon,* I pass through the same initial portal into the text, exactly as Joyce intends, with the narrator's wondering whether he had, unknowingly, seen his son die one morning. This time, by altering a single choice, choosing a path instead of a default, my reading experience changes again, shunting me onto different segments, new ones I had not encountered before. What I also stumble upon, however, is a relentless loop, where I shuffle through the same twenty-three segments repeatedly, the default from the twenty-third segment leading straight back to the first I had already read. My knowledge of the characters and events is enlarged somewhat, but I become disoriented by this circling through the hypertext, which spells the end of my second reading. When I return to the text for a third shot, I again rely on defaults to push me through it; however, I select a different place in the text to move via a path, and continue reading until the hypertext, again, refuses to default, handing me an excuse to pause—the way you might midway through, say, Walker Percy's *The Moviegoer,* not because you are tired of the narrative or tottering from so much evocative but understated prose, but simply to digest what you have read. This time, however, the narrative possesses more tensions and ambiguities than it did the first time around. Peter still has not discovered the whereabouts of his ex-wife and son; he may or may not be having an affair with a fellow employee named Nausicaa; and I am not entirely certain what kind of relationship binds him to Lolly, a sometime therapist who also happens to be the wife of his employer.

Usually, the further along my reading takes me through *The Scarlet Letter* or *The Long Goodbye,* the more the profusion of probable and plausible outcomes dwindle, declining from a torrent into a mere trickle of possibility until they are narrowed to a single outcome, a final conclusion. But, instead of narrowing the margins of the narrative the further I read, *afternoon* considerably broadens them. The more I read, the more contingencies arise, the larger the tangle of causal relationships grows. My third reading of *afternoon* provides me with still more inferences to verify, and I have, if anything, a less complete picture of the hypertext as a structure of possibilities than I had during my first reading. There is nothing in any of my three readings to satisfy either the sense of closure I know from years of reading both pulpy novels and canonized literature—let alone the definition of closure set out by the likes of Herrnstein-Smith, or even by Kermode and Brooks. There seems to be nothing whatever resembling a final, concluding metaphor in *afternoon* that organizes the patterns I have discovered in the text into a coherent, tangible whole.

By my fourth reading of *afternoon,* I become uncomfortably aware of mutually exclusive representations of events cropping up in each reading—most notably the lunchtime exchange between Peter and his employer, Wert, described earlier. In one version, the accident seems not to have occurred; in another, Wert distracts the worried Peter from his fears about the fates of ex-wife and child by making bawdy suggestions to their waitress. In one scenario, only Peter is having an affair with Nausicaa; in another, Wert knows both that he and Peter are having an affair with Nausicaa and that Peter is blissfully ignorant of Nausicaa's involvement with him. In one version of the scene, Wert idly wonders aloud how Peter would react if he, Wert, were sleeping with Peter's ex-wife, a childish ploy intended only to provoke Peter; in another, Wert is testing the extent of Peter's ignorance of his own involvement with her. While my readings of all these versions are physically possible, I cannot accept all of them simultaneously when I finally reach an understanding of the events described in *afternoon*— they represent the clashing of too many different worlds, too many readings of characters and events that are simply too disparate to be embraced in a single interpretation.

On my fourth reading of *afternoon*, my uncertainty about Nausicaa's involvement with both Wert and Peter is confirmed by a sequence of places narrated by Nausicaa herself. Most significantly, however, this particular version of the narrative rearranges the sequence in which Peter first sees the bodies of the child and woman splayed out on the green lawn. In this instance, Peter cannot track down either Lisa or Andrew prior to his driving to work, and his anxieties have already distracted him when he spots both of them riding in, of all places, Wert's truck. The possibility that Lisa may be sleeping with Wert—and perhaps, his recognition that Wert's lunchtime query may have been a very real question indeed—shocks him. Peter's feeling out of control is, in this version of *afternoon*, accompanied by his loss of control over his car. In an ironic twist, Peter *himself* causes the accident that injures or kills his wife and son—and it may be his feelings of guilt that prompt an amnesiac search for their whereabouts, one that both follows this sequence and begins most readings of *afternoon*. This reading ends, as did the first reading, on the place "I call," with the narrator relating his actions to us: "I take a pill and call Lolly"—only this time, he calls Lolly to assuage his guilt. And it is his calling Lolly that enables her to figure out, as we discover in the places "1/," "2/," and "white afternoon," that Peter has probably caused the accident.

To penetrate the narrative to its furthest extent, to realize most of its possibilities, I need, in a sense, to experience the place "I call" in each of the readings. The beginning of the therapy, introduced in my first reading of *afternoon* by the narrator's electing to call Lolly to stem his fears, becomes, through several encounters with the place "I call," an ongoing process of realization and discovery that culminates in Lolly's intercession, encountered in my last reading. It is this gesture of calling Lolly, in the end, that enables Peter to face the fact that he is culpable for the deaths or injuries of his ex-wife, Lisa, and son. As Joyce himself has noted: "In order to physically get to 'white afternoon,' you have to go through therapy with Lolly, the way Peter does," and it is only in the first and last readings that the place "I call" does not default, providing access to other segments at a mere click.[15] In all other readings, the segment defaults and also provides access to numerous other narrative strands. Of all the places in *afternoon,* "I call" has the largest number of paths branching out from it—ten— making it, significantly, a place both physically and literally central to the structure of the narrative.

What triggered my sense of having come to some closure, my sense that I did not need to continue reading *afternoon?* Most obviously, I became conscious of my readings having satisfied one of the primary quests outlined in the narrative: what has happened to Peter's ex-wife and child? Although my discovery that Peter has caused the accident is not entirely congruent with his desire to learn of their condition (unless, of course, he's suffering from either amnesia or denial), it does short-circuit Peter's quest. Since Peter himself has caused the accident, clearly, he knows whether the pair is unharmed, fatally injured, or already dead. The language in the place "white afternoon" suggests the last possibility may be the most valid: "The investigator finds him to be at fault. He is shocked to see the body . . . on the wide green lawn. The boy is nearby." The word "body" may signify that the woman Peter sees is lifeless, but it could also refer to the fact that she is unconscious, inert, quantifiable as an accident victim. Although he does not identify the bodies he sees in this segment, elsewhere in the narrative the absences of both Lisa and Andrew from home, office, and school suggest that they might be the accident victims Peter sees. Further, when Peter revisits the scene of the accident, he comes upon crumpled school papers written by his son, which may have fallen out of one of the vehicles on impact, and is moved to tears—again strongly suggesting that he has caused a fatal accident.

So I can, to borrow the expression from the print world, close the book on *afternoon:* I am finished with it. I might pick it up in a few months or years' time, the way I could watch *Chinatown* a good five or six times in a decade without exhausting my pleasure in the way the narrative simultaneously evokes, includes, and demolishes the whole world of film noir—but, in any case, these types of returns are nearly inevitably prompted by satisfaction with your aesthetic experience, not from a nagging sense of incompleteness. At the same time, I cannot help but be aware of the mutability of Joyce's narrative, the way it appears to shift, chameleon-like, each time I open it, becoming, sometimes, a story about everything *but* Peter's quest to locate his wife and son. I am not, for example, absolutely certain that Peter didn't simply see his ex-wife keeping company with his employer, swerve and strike another car, carrying an unknown woman and child in it. That would certainly account for the "investigator finds him at fault" line, as well as the bodies stretched out on the grass, but not his son's school paper, blowing about on the grass—just as it would also leave Peter's search for Lisa and Andrew as open-ended as it was when I first began reading *afternoon.* Which makes all the more intriguing the reasons for my closing *afternoon,* feeling satisfied with the last version of the text I read, and accepting the approximate, albeit stylized, type of closure I reached at that last "I call."

First, *the text does not default, requiring that I physically alter my reading strategy or stop reading.* Since the segment "I call" also refused to default the first time I encountered it, what distinguishes my first and last experiences of this physical cue? Why does it prompt me, the first time I come across it, to read the narrative again from the beginning, pursuing different connections, yet prompt me to stop reading the second time? The decision to continue reading after my first encounter with "I call" reflected my awareness that the first reading of *afternoon* visited only 36 places out of a total of 539—leaving the bulk of the narrative places still to be discovered on subsequent readings. And, further, on the first reading, I perceived the failure of the text initially to default from the place "I call" as an invitation to return to the narrative. This recalls the redirection of textual energies that Brooks mentions in his analysis of Freud's narrative of the Wolf Man:

> Causation can work backward as well as forward since the effect of an event . . . often comes only when it takes on meaning. . . . Chronological sequence may not settle the issue of cause: events may gain traumatic significance by deferred action or retroaction, action working in reverse sequence to create a meaning that did not previously exist.[16]

That is, this physical "conclusion" to the narrative sends me back into its midst to discover the cause behind Peter's anxiety and to resolve additional questions that my journey through the narrative has already raised. Readers of Freud's narrative about the Wolf Man may have to page backward to assemble their own versions of the causation and motivation behind the occurrences they have discovered in the narrative. But on the other hand, as a reader of a hypertext narrative, I am fairly certain that further readings of *afternoon* will yield a different chronology, different apparent motivations, and even a different set of events leading to a conclusion totally dissimilar to that of the narrator gulping a pill and reaching for the phone to call his therapist friend.

Second, *this particular conclusion represented a resolution of the tensions that, initially, give rise to the narrative.* The fiction begins with two quests: Peter's search for the whereabouts of his ex-wife and son, to confirm whether they might have been the accident victims he glimpsed that morning, and our seeking a better sense of exactly what it is that Peter saw on his way to work. When we look at the accident through Peter's eyes, we see only the scene revisited by him several hours later (when, in any case, there is practically nothing to observe)—and we cannot begin to account for his nearly paralyzing fear that the bodies he saw so briefly might belong to those closest to him. It is also more than simply an irrational tic, since, for all Peter knows, his son might be busy teasing a girl in his class and his ex-wife slogging through a thoroughly average business day even as he's dialing his index finger into a bruised stub. There are no absences to provoke his fear, and the mere proximity of the accident to his son's school does not account for the crescendoing panic driving his actions. Nor should the tone of the perfectly banal conversations he conducts with people who cannot recall whether they have recently seen Lisa and Andrew, unharmed and going about their regular business. My sense of the significance of "white afternoon" lies partially in its ability to account for the undertone of hysteria edging Peter's fear. If Peter has caused the accident that has injured them but has blocked this horrifying bit of knowledge from his consciousness, his inquiries would probably have this particular character of concern mixed with panic. Put another way, Peter's panic-stricken inquiries and fearful conclusions do not match any script I can recall from either experience or from other narratives that describe a search for the whereabouts of missing family members or friends. His emotions do, however, correspond to scripts familiar to me from other stories or films. In *Angel Heart*, for example, the Mickey Rourke character frantically tries to discover the identity of the killer trailing in his wake, who seems to be following just scant steps behind him, only to find that the murderer is not merely right behind him—he *is* him.

Third, *the conclusion represented a resolution that accounted for the greatest number of ambiguities in the narrative.* In other words, this version provided the most *plausible* conclusion to the narrative's network of mysteries and tensions. Just as plausibility and referentiality bind sentences and paragraphs together like invisible glue, inviting us to see logic, causation, and sequences linking lines of text together like a chain, they also prejudice me toward certain conclusions. The conclusion that refers to the most narrative tensions, that stands for the most plausible model of the motives and intentions of characters like Peter and Wert, seems the most satisfying. Wert's romance with

Lisa accounts for the peculiar tenor of some of his comments to Peter, but it also coincides with the way Lisa, Lolly, and Nausicaa, throughout the narrative, bemoan Peter's peculiar proclivity for imagining himself at the center of everyone's universe. Also, I cannot seem to account for the contents of "white afternoon" through any other way of interpreting the narratives in *afternoon,* regardless of how artfully I make my guesses about the pronoun referents. If Peter does not crash his car (and into Lisa and Andrew) on his way to work that morning because he realizes his womanizing boss has taken to sleeping with his ex-wife, then how should I construe Lolly and Nausicaa's concluding that they should not blame "either of them" as they mull over the accident? This reading of "white afternoon" also accounts for the otherwise puzzling places "1/" and "2/" in Lolly's monologue:

> Let's agree that it is shocking, unexpected, to see this particular woman with [Wert]. Yes, I know that, for anyone else this should not be unexpected, that Peter should, at least, have suspected; but we nonetheless ought to grant him his truth. It is all he has, and so it is authentic. Let's agree he must feel abandoned—even, literally, out of control.[17]

> Wert knows Peter takes this road.
> Peter knows we women are free. . . .
> The world is a world of properties and physical objects, of entropy. . . . Even coincidence is a free-will decision.[18]

So, the "either of them" in "white afternoon" must be Peter and Wert: Peter for being so self-centered that the shock of seeing his still beloved ex-wife with another man causes him to swerve across the yellow line; Wert for deliberately, perhaps maliciously, driving along a road where Peter will probably spot them—a gesture not unlike the old "What if I were sleeping with your ex-wife?" query tossed out between courses at lunch. Read this way, the string of segments that culminate in "white afternoon" and "I call" have all the coherence of a cinematic sequence that spans climax and denouement, plugs all the missing gaps, and overturns my hypotheses, so to speak, in the same way that the flashback sequence toward the end of *The Other* jolts me with the realization that the good twin, who, to all appearances spends nearly the entire film spinning in the wake of his evil brother's destructive acts, really is himself the bad twin.

Its murky spots and ambiguities resolved by this interpretation, the narrative suddenly looms into focus as a unified whole, a structure of possibilities representing one man's simultaneous drive to learn the fates of his ex-wife and son—as well as a mad dash away from his own culpability in an accident that may have caused their violent ends. If the narrative pushes us to follow the paths Andrew and Lisa might have taken to discover their fates, it also prods us to uncover truths Peter himself is too self-absorbed, insecure, or downright terrified to admit. So Lolly's monologue, ending in the revelation that Peter has caused the accident, is, in a sense, the destination toward which the narrative has been pushing us from two entirely different directions, baring some of the tidbits Peter could never bear to admit even to himself about his boss, about his former wife, about his own shortcomings, at the same time that it reveals what has happened to his family. Once I have reached it, I am able to shuffle back through my experience of the entire narrative and to see it as a chronicle of Peter's ongoing denial of everything from his feelings for his ex-wife to his role in the car accident. In other words, I reach a point where I perceive the "structure of the work as, at once, both dynamic and whole"—which, coincidentally, manages to neatly satisfy Herrnstein-Smith's definition of conventional narrative closure.[19]

Fourth, *my interpretation of the significance of "white afternoon" is tied to my perception of "I call" as a central junction in the structure of the text and of "white afternoon" as a peripheral, deeply embedded, and relatively inaccessible place.* Joyce himself is the first to point out that the cognitive map of *afternoon* reflects his organization of the narrative as he wrote it and not the structure of readers' potential encounters with it.[20] But this does not prevent me from discovering striking concurrences between my perception of the virtual space occupied by segments such as "I call," and "white afternoon," in the overall structure of the hypertext and their position on the cognitive map of *afternoon*. The narrative's network of guardfields, requiring readers to visit a particular segment or select a certain word or phrase from its text, appears to track readers through the hypertext in a highly controlled manner—witness the way I could realize consistently different versions of the narrative by simply making a change at one point in my navigations. In particular, my ability to visit certain portions of the hypertext seemed largely contingent on whether I had visited "I call" and how frequently I had been there. The sequence of places visited tracks me through the text via these conditional links, making certain paths accessible and certain defaults tangible, causing my experience of the text to somewhat resemble Dante's penetration of the rings of Hell in *The Inferno*. The more I read the narrative, the closer I approach to its center—and, like Dante, I cannot suddenly emerge in the environs known to Judas Iscariot in the very pit of Hell without having first visited the more lofty realms populated by those who merely lived lives without benefit of Christian baptism. In order to reach the pit where Peter becomes his ex-wife's (probably inadvertent) executioner, I first have to trundle through a narrative that lets me conclude that he might very well have been a lousy husband, several rings up from the bottom.

The place "I call" seems to exist as *afternoon*'s central junction, where readers are switched onto certain narrative strands that spiral down further into the narrative with each successive encounter. Significantly, the place "white afternoon," along with the rest of the sequence revealed in Lolly's monologue, is embedded at the deepest structural level of *afternoon*, five notches below the uppermost layer of the narrative, the one through which readers first enter the text. Only two connections lead into this narrative strand, and a succession of guardfields ensures that it is reached only after a lengthy visitation of fifty-seven narrative places. Hence, my sense, when I get there, of arriving at the end of something, because "white afternoon" represents the furthest reaches of the physical spaces within *afternoon*, the textual equivalent of a basement—or the end of a novel.

The Suspension of Closure:
WOE—or a Memory of What Will Be

> It is a story of being at the edge of something. That is not
> authorial intention but discovery. If in doubt how to read, ask
> your teacher or your heart.
> —Michael Joyce, *WOE* (1991)

We could call *afternoon*, with all its layers of text and levels of meaning, "stratigraphic" writing. Once we open it, we can delve through layer after layer of singular versions of narrative events until we reach the bottom band, the one that contains the deepest, most secret record, in the same way the oldest stratum in sedimentary stone tells geologists tales of what lived and died in the shadowy recesses of the past. The trigger that sparks off Peter's frenetic quest, encased in Lolly's narrative, slumbers on the bottom level of *afternoon*, shaping everything that follows it, including Peter's denial of all knowledge of what might have happened on his way to work. Although *afternoon* is not a mystery in the conventional sense, its action takes its central thrust from that genre's narrative dialectic of concealment and discovery that drives events forward in nearly every narrative strand. Peter's quest makes me itch to fill in the fine details. Once I can sketch them in, however, the whole search is, essentially, up—even if *afternoon* lacks the traditional denouement beloved of Victorian novelists and Hollywood moguls alike, where the perpetrator either ends up straitjacketed and muttering to himself in an institution, enveloped in the bosom of his family, or behind bars. Seen in this light, it is not terribly surprising that the narrative should prompt and finally satisfy my search for a rough equivalent of narrative closure—albeit a search somewhat satisfied through avenues beyond the boundaries of print narratives.

On the other hand, this sense of ending-ness could just be unique to *afternoon,* making closure something that hypertext narratives can, but may not, have. Even if *afternoon* is, as Stuart Moulthrop has argued, "only a 'mystery' in the older sense of that word, the sense of ritual or hieratic procedure,"[21] it is a narrative that begins with a question that begs to be answered: a matter of life and death. It is difficult to imagine a more potent formula to compel readers through a text, even a network as dense and circuitous as *afternoon.* As a genre, mysteries could have been designed for hypertext or hypermedia—as writers like Shannon Gilligan have clearly already discovered. The whole pleasure bound up with the reading of them revolves around the narrative leaving a vacuum that invites us to hurl any number of hypotheses and hunches into it. But whereas Raymond Chandler or Sara Paretsky eventually narrows the field to a single culprit with a single outcome conferring on us the fleeting pleasure of knowing we had called all the right shots and interpreted the signs as well as our hero did, a digital narrative like Gilligan's *Magic Death* urges us to go on guessing, reminding us that life, like texts, is indeterminate. The case can always wind up differently, the guilty party turn out to be somebody else: the sweetly inquisitive elderly neighbor, the brother honking his grief into a handkerchief so convincingly during questioning, the ones you would never suspect. So the questions remain: if I were to read the hypertext equivalent of a melodrama or something distinctly Chekhovian that does not spur me on, panting after the answers to a few pressing questions, would I still read for closure? Or would closure become relatively unimportant? If it does, then would it be possible for me to read these narratives comprehensibly? And how on earth would I figure out when (and where) to stop? In a text that has no rending narrative tensions, will I discard my search for resolutions? Or will I impose or even invent some, to confer some shred of purposiveness on my readings?

Like Robbe-Grillet's *In the Labyrinth,* James Joyce's *Ulysses,* or Virginia Woolf's *Mrs Dalloway,* Michael Joyce's hypertext, *WOE—or a Memory of What Will Be,* is a narrative "about" its own structure—a radical notion even during the modernist era. On the face of it, the Joyce and Woolf novels simply span a single day in the lives of two unlikely pairs: Leopold Bloom and Stephen Daedalus, Clarissa Dalloway and Septimus Warren Smith. Both of the days involved, however, are loaded with recurrent patterns of tensions, conflicts, meditations, and ambiguities that extend their tentacles both into the distant past and forward into the future, so that few of them are resolved by the closing of the day.

The plot of *Mrs Dalloway* has at least one overt question: what will happen to Septimus Warren Smith, who never entirely came back from the battlefield (for all he manages to stumble through the appearances of normality)? And the novel certainly dispatches with it, quite permanently, when he hurls himself out a window and ends up impaled on the iron railings lining the pavement below. This, nonetheless, occupies only a fragment of the narrative, which goes on to trace the origins of regrets, ambitions, desires, and decisions that drift through the minds and memories of Clarissa, Hugh, Richard, Peter Walsh, and Lucrezia Warren Smith. We expect the two parallel narrative strands involving the very different days spent by Clarissa and Septimus Warren Smith to intersect, and for their collision to alter the trajectory of both, but we discover that their lives glide by one another in perfectly parallel lines. The nearest Clarissa approaches the life of Septimus Warren Smith is one of the eddies cast out by his suicide, when the doctor invited to her party is detained by Smith's death, and the ambulance wailing down Tottenham Court Road on its way either to or from the place where the dying Septimus lies merely interrupts Peter Walsh's thoughts of Clarissa.

Like *In the Labyrinth* and *The Good Soldier, Mrs Dalloway* is structured around what Joseph Frank dubbed "spatial form."[22] Modernist literary works, Frank noticed, attempted to convey simultaneity and the often circuitous and incomplete patterns of fleeting thoughts through literary devices that included recurrent images, fragmented narrative sequences, and the division of plot from narrative—patterns that acquired significance, he believed, when perceived as part of a whole in the minds of their readers. A fully fledged interpretation of them, the revelation of their full meaning, could occur only after readers had finished reading the entire text. In this view, meaning resides not simply in an ending that either confirms or subverts the expectations the narrative has fostered, but in the relationship between the content of the text and the place it occupies relative to the work as a whole. Taking Frank's concept further, David Mickelsen has argued that novels employing spatial form "are far from resolved" and are, instead, open works formed largely as explorations:

> The world portrayed is in a sense unfinished (unorganized), requiring the reader's collaboration and involvement, his interpretation. . . . [T]he "implied reader," in Iser's phrase, in spatial form is more active, perhaps even more sophisticated, than that implied by most traditional fiction.[23]

Obviously, the spatial form at work in these print narratives exists in our minds—like the maps of the structure of "Forking Paths" that the cartographic readers created—as we grapple with their intricacies of time and place, with patterns of recursion, and with digressions that violate expectations based on readings of conventional narratives. When a character like John Dowell tells his story in an order dictated by all the vagaries of memory, when Borges prefaces "The Garden of Forking Paths" with an allusion to an historical event seemingly unrelated to Yu's lengthy confession (itself discovered in fragments, its earliest pages missing), or when Justine goes looking for her kidnapped child in a brothel in *The Alexandria Quartet* and the novel reveals, nearly seven hundred pages (and three whole books) later, that she has discovered its final resting place, fiction requires us to attend to spatial form. Even in more straightforward, conventional texts, the slightest distance between time as it occurs in the plot chronology and time as it is traced by the narrative requires us to piece together intricate chronologies in our heads, and to assign significance and weight to events on the basis of where they appear in the narrative and the places they occupy in plot-time.

Although it may seem difficult to think about space in narratives that are, after all, made of lines of text laid out on flat pages, we inevitably think about their contents along two axes: time and space. The works that appear to deal with "spatial form" are those that insist we consciously and constantly deal with both dimensions. An author's aim might be to make time as visible, as palpable as space, as Proust attempted in *Remembrance of Things Past*—right down to his original notion of assigning sections of the narrative to spaces in a cathedral: "Porch," "Stained Glass of the Apse."[24] Authors like Woolf might try to capture simultaneity, a single event perceived from multiple viewpoints, as when the mysterious car glides across London, inflaming speculation about its occupants, and Woolf's narrative skips lightly between the consciousnesses of the observers who watch it pass. Or, as in "The Babysitter," a single occurrence might belong to entirely different scenarios, like the babysitter's scream, which metamorphoses from a squeal of terror to an indignant shriek, depending on the context we see it in.

Just as Iser reminds us that texts are only skeletal maps until readers flesh them out, so critics focusing on spatial form insist that it exists only in latent form on the page. Readers actually create it as they shuffle bits of text around in their heads, trying to get a bead on the layers of narrative time, juxtaposed images, recurrent themes, multiple perspectives on events, and even parallel lives: "Verbal space acquires consistency as the stylistic rendering of the text becomes apparent: reiteration, allusion, parallelism, and contrast relate some parts of the narration to others, and the construction imposes itself on the reader through the action constituted by the reading."[25]

In "Spatial Form in Modern Literature," the treatise that first brought the concept of spatial form to the attention of critics, Frank claimed that readers exploring narratives that use spatial form were required, by the very nature of this pattern of references, ellipses, recursions, and fluctuating points of view, to suspend "the process of individual reference temporarily" until completing the narrative, when "the entire pattern of internal references can be apprehended as a unity."[26] This particular method of reading a narrative as a structure or pattern of references is hardly a novel concept since, as we have already seen, many critics believe our perceiving patterns and forging hypotheses about their significance draws us through the narrative, anchoring body to conclusion. But for Frank, we do not simply require an ending to ratify our guesses and assign a definite and fixed value to what we have read—we can perceive the full meaning of complex, spatial works only when we reencounter them again, piecing the whole thing back together in our minds, or, even, reading the entire work over again. This conviction is reduced to a more elegant formula in Frank's famous declaration: "Joyce cannot be read—he can only be reread" (91).

It is not terribly difficult to accept that we can grasp the full meaning of *Ulysses* or *The Good Soldier* only once we have finished reading them and reflect back on their jumbled chronologies, patterns of reference, and hiccups in time. It is another, however, to contend that readers can suspend the act of creating or construing references for the words they read until they can place them within a global vision of the text. While we may leave some of our assumptions open to future revision, we cannot surge forward through any text without making assumptions. Without making inferences, without construing references and significance, we cannot read—and this is one of the few aspects of reading on which most theorists and critics agree. Part of the problem with Frank's theory lies, as critics like W. J. T. Mitchell have pointed out, in his insistence that spatial form is the property only of "modern avant-garde writing."[27] The concept of spatial form, like Frank's essay, is useful for its reminding us that texts occupy as well as describe both time and space. As a realistic model of how readers approach works by Pound, Eliot, or Joyce, though, Frank's theory flies in the face of the admittedly modest amount of wisdom we possess on the act of reading.

Even as we begin plowing through our first sentences in, say, *Ulysses* or even the likes of *Princess Daisy*, we are busily interpreting, integrating details, concocting hypotheses, modifying, confirming, and abandoning predictions.[28] The "glue" binding disparate elements spread out over several hundred pages is our ability to perceive references between sentences and paragraphs: to see all the pronouns anchored to precedents and to see sentences aligned in chains describing causes and their effects. The very act of reading requires us, albeit generally unconsciously, to continually perceive links, references, and contexts for the words we read, which come to us already endowed with meanings at the moment we perceive them. "Meanings come already calculated," Stanley Fish has famously argued, "not because of norms embedded in language but because language is always perceived, from the very first, within a structure of norms . . . [a] structure, however, [that] is not abstract and independent but social."[29]

Far from the passive consumers Frank envisioned, who merely assemble pieces and refrain from assigning them a value until the whole has been revealed, Fish's readers construct as much as they construe. The readers of "Forking Paths" and "The Garden of Forking Paths" interpreted the narrative according to models of the integral structure they had erected, as when the readers of Borges's short story insisted that its conclusion betrayed the entire notion of the labyrinth. In "How to Recognize a Poem When You See One," Fish's readers, believing names left over on a blackboard from an earlier linguistics class were part of an esoteric piece of poetry, constructed a poem from them, based on the significance they assigned each name and its alignment on the board.[30] Even in texts where both story and narrative grow from a complex network of recurrent themes, densely interwoven thickets of time, and clusters of multiple perspectives, we do not suspend the action of construing/constructing, as Frank insists. What seems more likely is that we are unable to form determinate predictions, as we tend to in our readings of narratives with clear-cut conflicts and tensions calling for tangible resolutions. Instead, our coming to closure on these spatial or exploratory narratives involves our ability to construct models of the narrative structure that assign a place, weight, and significance to the associations and themes we have encountered—an action that recalls the efforts of the readers navigating through "Forking Paths," as well as my own readings of *afternoon.*

In other words, as readers we have always been challenged with the task of reading something that approximates the virtual, three-dimensional space of hypertext narratives: what Frank insists is the hallmark of the modern novel seems to have been the property of fiction since the days of Richardson and Defoe. What does lend a ring of truth to Frank's declaration that *Ulysses* can only be reread is perhaps Joyce's wrestling with what Jay Bolter calls spatial or "topographic" writing in a unidimensional, undeniably static medium.[31] To understand *Ulysses,* we need to construct the equivalent of Greek rhetoricians' detailed memory palaces in our heads, where we can weigh narrative-time against story-time, Joyce's narrative against the familiar conventions of the novel. To understand the ellipses, leaps in perspective, and disjunctions in time in Michael Joyce's *WOE,* however, I need only crank open the cognitive map and peer into the structure of the hypertext.

Indeterminacy, Spatial Form, and WOE

WOE opens with a place called "Mandala," a segment of text that lies over a cognitive map of the hypertext that itself looks like a mandala. "Mandala" represents the hub of a narrative wheel, connected through a series of paths to five other places that, in turn, contain other, subsidiary places (see fig. 1). From "Mandala," however, I need not pass through the five places on the uppermost layer of the narrative in order to gain access to the levels of narrative within each of these five places: a series of links, paths, and defaults connects some of the text's most embedded places with "Mandala," the point of entry into *WOE.* In Buddhist practice, the mandala pulls the eye from the center of the image to the periphery or vice versa. In *WOE*—its title a pun on the acronym of the journal *Writing on the Edge,* for which the piece was written—the map is also a visual pun and a metaphor for a form of writing on the edge.[32] Physically, the reading of all places except "Mandala" takes place at the periphery, or on the edge, of the narrative's structure, making the narrative itself, quite literally, a writing on the margins of experience, an accumulation of the experiences, memories, and metaphors from which fiction grows.

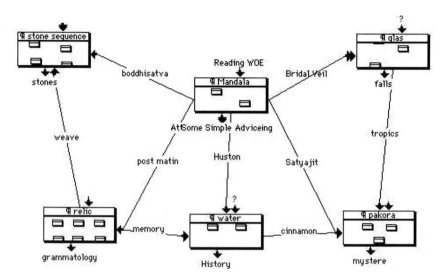

Fig. 1. The cognitive map of *WOE—Or What Will Be:* writing and reading at the edges.

Although my reading of "Mandala" is colored by my awareness of its central place in the *WOE* narrative structure, my knowledge of its placement at the hub of the narrative structure does little to relieve the ambiguities littering its text. Here the protagonists are identified solely by the pronouns "he" and "she," and the pair seems to be driving somewhere, but I cannot be certain even whether their journey is a physical one or merely metaphoric. As I move through the narrative by way of defaults, I encounter more scenes that portray the actions of a couple, but I cannot be certain whether these two pronouns are the same ones I passed earlier, taking what may or may not have been a trip in the car. When we read print stories, if we lose track of which name is "he" or stumble over the identity of the next "she," we simply trace the text backward to verify the pronoun referents, until we know that "she" is Leonora or Clarissa or Lucrezia. With *WOE*, however, tangible gaps separate pronouns from precedents, and even when I traipse through the text via its defaults, which in Joyce's hyperfiction usually string the text together in sequences, I cannot assume comfortably that the "she" in one place is the same "she" I meet another three places down the line. Because I also know that the same place can crop up in wildly different contexts from my experience with *afternoon,* I am distinctly aware of my hesitation as I cast around for the likeliest anchor for the pronouns.

Of course, the central story in *WOE* also has something to do with my treading so carefully: two married couples, one identified solely through pronouns, joined through friendship and infidelity. Sometimes the "he" and "she" are labeled "Steve" and "Filly"; sometimes they are simply "she" and "he." Often, the pronouns refer not to this couple but to their friends, who can be distinguished from Filly and Steve only when their children are mentioned (since Filly is unable to have children) or when the text makes it clear that they are *not* Filly and Steve. The titles of the places involved in the narrative sequence "Relic" that focuses on the couples are identified by pronoun titles— "She," "They," "He," "It," "Your," "Their," "His," "Her," and "We"—almost a teasing promise that this jumbled network of identities will be smoothed out in each place, all the female pronouns tamed into a single referent, the respective characters that make up any given "we" pinned with bright identity tags. Since I quickly learn that one "he" is sleeping with the "she" with whom he did not swap wedding vows, my reading of *WOE* develops into a quest to establish and fix the identities of these wayward pronouns. However, because pronouns in hypertext narratives can be more slippery or promiscuous than any poststructuralist theorist ever dreamed, even my wariness cannot prevent me from unknowingly making some determinate assumptions.

In "Their," a man and woman converse. The woman must be the one with children, since the man here wonders if she is wearing a certain fragrance because she has realized he loves the scent of Filly's perfume. That leaves me to establish the identity of the man, but, as I begin reading the passage, he remains elusive, a set of XY chromosomes, Anyman . . . until I read that the two of them have packed the kids off to see the film *Dick Tracy.* Sounds like the sort of thing a couple seeking a little privacy or some quasi-spontaneous sex might do, a familiar enough schema. On the other hand, if the couples are such good friends, neither the children nor the unnamed woman would think it amiss if Steve—the man who *is not* her husband—were to ply her kids with dollars in an avuncular way and send them off for an afternoon at the cinema, so the two of them could spend a few hours in the sack together, another familiar and highly plausible schema. I continue reading, and it is only when the man reacts with shock when the woman tells him she believes he's thinking of Filly that I latch onto a certain sense of his identity. His silently wondering "Do you know?" seems motivated by guilt: Why else would he feel a shock, if he is not thinking of another woman while he lies alongside his wife?

What does he know that plainly she does not? There are enough of its trappings here for me to recognize the old "adultery" script, familiar to me from my encounters with print and film narratives about ménages à trois. So, I believe, the man here must be involved with his wife's friend Filly, although the text does not make explicit just how much the wife knows. Since the schema or script for adulterous relationships invariably involves a dialectic between deceit and discovery, however, the question of the wife's knowledge or ignorance of the affair is one of the engines that keeps the narrative—not to mention the affair itself—barreling along.

When I begin reading the place "His," which immediately follows "Their" by default, I assume that the "she" embraced in "his" arms in bed must be his wife, the same "she" I ran across seconds before. But I am jolted when Steve interrupts the couple's postcoital musing by leaving a message on "his" answering machine. Since Steve cannot be both leaving messages on answering machines and fondling someone in bed at the same time (and in most instances people do not call their own answering machines to leave messages on them), I realize that, although the man is the same, the woman may not be. This "she" is startled by the messages left by Steve, who is identified as her husband, so she must be Filly, Steve's wife. Although the "he" lying in bed might be a third man, Joyce's narrative seems to cleave close enough to the conventions of print narratives for me to believe that new characters would be introduced with some modicum of fanfare, some indication of their debut. Still, this abrupt switch in identities comes as a shock, even though it is couched in an environment that has encouraged me to believe I can leave behind all the baggage of assumptions, projections, and conventions I bring to my experiences with the printed word. Mere continuity, supplied by the default connection between "His" and "Their," has led me to assume, as I would in print narratives, that the actors in both places will remain constant, making my surprise at the switch in identities all the more potent. Here the gap separating narrative spaces approximates the space of cinematic cuts—which, although we know at some level are splices in the film, we still believe smoothly connect one image to another. My reaction is similar to the experience of someone watching adjoining scenes in a film about an affair, where in two separate sequences the slow pan of the camera moving up the intertwined bodies of a man and woman reveals two different women's faces topping seemingly identical sets of thighs, hips, breasts.

I read on, expecting the text to seem more determinate, since my predictions about the deceit/discovery dialectic should encourage me to see language in a more determinate, meaningful context. I find correspondences between another narrative strand, involving yet another unnamed woman who murders her philandering spouse before killing herself, and the ménage à trois in "Relic," which may foreshadow the violence with which the nameless wife may react once she discovers her own husband has been sleeping with her best friend. Physically, the tale of this double killing resides at an entirely different subsidiary level of *WOE*, but the connections I make between these two distinct narrative strands easily bridge the gaps between them, a feat, apparently, many readers accomplish with ease:

> Recent extensions of the concept of macrostructure suggest
> . . . that the macrostructural hierarchy is also "networked":
> the repetition in a text of a previously mentioned element
> may form a connection between the two related propositions,
> even if they are at different branches in the hierarchical
> macrostructure. . . . The macrostructures which readers build
> of texts allow them to organize and reduce complex informa-
> tion to a meaningful, manageable whole.[33]

The familiar script of adulterous couples encourages me to forge certain predictions about how the narrative will develop, how the tale of Filly and Hubby's stolen afternoons might blossom into a story of a rage unleashed, marked by a trail of bodies, a schema into which the double murder fits quite nicely, and I take it as a temporary confirmation of my hypothesis. As I read on, however, I keep running up against segments that seem to have no bearing whatsoever on the events in "Relic," let alone any answers to my questions: Does she or doesn't she know? Will she act, or won't she? Because my reading is now purposive, turning *WOE* into a thriller that has a clearly defined set of oppositions, I gloss over some of the same indeterminacies that had earlier excited my attention in places like "His" and "Their." In any case, I have no other context against which to set the contents of these other places, so I simply assign this newer information a place in the background to my reading of *WOE* while I pursue further developments with the "Relic" couples by consulting the cognitive map to fix their whereabouts.

What I discover momentarily shocks me. "Relic," I had believed, was the axis of the hypertext, uniting its disparate narratives with themes of impending violence, violated fidelity, and bonds forged by love and desire. According to this view, "Relic" should have occupied the center of *WOE*, or something close to the position of "Mandala,"

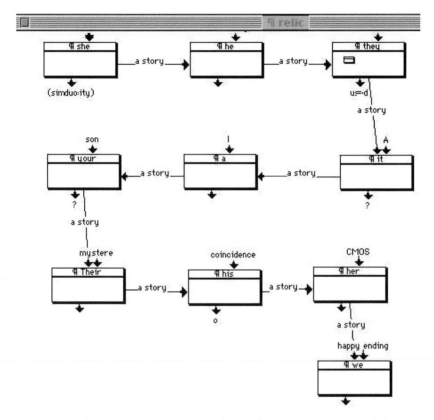

Fig. 2. The cognitive map of "Relic" tells its own version of the story.

with other narrative strands feeding into its motifs, enlarging upon them, foreshadowing the future in "Relic." Instead, I discover, "Relic" is merely one of five places on the periphery of *WOE*. Although, of course, I had noticed the mandala-like shape of *WOE* when the hypertext opened into both "Mandala" and its cognitive map, I lacked any context to make this knowledge meaningful, and, as we tend to do with things we perceive that do not seem particularly relevant at the time we first notice them, I pushed these details to one side. Now, confronted with them again, I realize the couplings and uncouplings of "Relic" cannot be as central to *WOE* as I had assumed (see fig. 2).

The words "A happy ending" conclude the text of the place "We." But when I first see them, I read this phrase ironically and attribute the words to the unnamed narrator of this particular passage, possibly the child of the married couple, who lists the name of his beloved family members but leaves out the name Liam, which must, therefore, be his own. If Liam is the "me" in this passage, and he is reporting things in what must be a tense household, then, like the child in *What Maisie Knew*, his "happy ending" is most likely going to be anything but that. Far from seeming an ending, these words originally appear to be a way of heightening tension—one of those familiar tricks like the moments in *Sleeping with the Enemy* and *Misery* when it looks like the heavies have gone down for the count, inviting you to relax and heave a sigh of relief . . . before the narrative propels them upward again, Rasputin-like, for one last tussle with the horrified protagonists. The cognitive map, however, tells me otherwise. According to it, I have already experienced all the twitchings and couplings "Relic" contains, a relatively short sequence that ends with "We." Even the connection linking "We" with the place that precedes it is named, echoing the text, "happy ending," obliging me to revise my assumption that this place merely marked the narrative's downturn into a more violent phase. The child's phrase "a happy ending" should, after all, be read without irony, a simple statement of fact: this particular story is finished. The tensions I discovered in the narrative seem to have existed in my imagination more than on the virtual pages of *WOE*—and I may well have drawn upon my knowledge of familiar schema about adultery and revenge, however subconsciously, to make my reading of *WOE* more quest oriented and, therefore, easier.

At the same time, I have to admit that the places in "Relic," as the most decidedly sequential in *WOE*, also seem to draw the trajectory of the rest of the narrative into them by a kind of centrifugal force. It is hardly surprising, from this perspective, that I felt prompted to read so many of the other, disparate places in the narrative in light of their real or imagined references to, and consonances with, the details contained within the "Relic" narrative. But if I view the story in "Relic" as limited to what I have already read, the avenging wife and her dead husband become something other than portents of things yet to unfold in the story of Filly and Steve and their friends. Approached again, however, these events seem to be related in some way to Joyce's own past: involving a cousin who murdered, then killed herself, leaving behind an orphaned child—particularly when read against fragments of what seem to be Joyce's diaries or a journal, as well as metatextual musings on the act of writing *WOE*. By carefully comparing the places I read in *WOE* with their position on its topographic map, I eventually evolve a sense of *WOE* as a narrative about the creation of hypertext fiction, brewed from snippets of experience from Joyce's own past in diary-like, dated extracts, fragments of the experience of others, scatterings of news items and poetry, and metatextual commentary on the act of creating *WOE* itself—each representing one of the places ringing "Mandala." This revised sense of the narrative structure of *WOE* grows slowly, as I create a network of references and connections between places—much as readers of *Ulysses* or "Forking Paths" might read. If closure is a sense of completion, though, surely it cannot look like this: I feel far from having resolved many of the ambiguities hanging around the text. Did the double murder "really" happen? Is this nameless woman who murdered her cheating husband the same woman, "M's sister," who I discover has been murdered? Does the wife ever find out her husband has been screwing around on her? Does Filly's husband discover his best pal's duplicity—or do the friendships remain intact, the affair undiscovered?

Despite this, I feel as if I have completed *WOE*, discovered a level place from which I can, a little like Archimedes with his place to stand and his fulcrum, grasp the world—in this instance, the text, its narrative structure, and some of its imagery and thematic references—and fashion them into a plausible reading that accounts for the majority of its form and content. This is, perhaps, a brand of closure akin to the sense of an ending you might arrive at after reading *Ulysses* or *Finnegans Wake*, a feeling you have merely completed one plausible reading of the text, realized one version without exhausting the many others still possible, still to be discovered. Perhaps this sense of a completion is informed less by what I have learned about the structure of *WOE* than it is about my knowledge of other print narratives, like John Barth's "Lost in the Funhouse," which are also mostly "about" the relationship between narrative structure and the experience of reading. In Barth's short story, the Funhouse is both a physical place—tricked out with all the usual halls of mirrors—and a metaphor for our own readerly way of getting "lost" amid all the mirrors and false turns and fake walls thrown up by fiction.

But my arrival at a sense of completion for *WOE* also represents the kind of strategy by which closure can act as an elegant form of shorthand, a way of providing a single interpretation of events that most efficiently accounts for the greatest amount of information. Had I attempted to find a metaphor or attach significance to each of the disparate texts I encountered in *WOE*, I would still be at it. Once I decide, though, that Joyce's hypertext is about the production of coherent, neat narratives from the inconclusive, fragmentary flotsam of everyday life, this reading accommodates nearly all of these passages and effectively neutralizes their disparities. Of course, I may be more inclined to leave the loose ends in *WOE* flapping than I was with *afternoon*, because part of what *WOE* is about, I suspect, is the flexibility of certain works of art that can invite readers and viewers inside their framework repeatedly for fresh runs at involvement and interpretation without ever exhausting the work itself. I can close *WOE* without worrying what my reading neglected, since I realize that, as an Open Work, no single version of the text can truly complete it.

Reading for the Ending: Closure in Print and Interactive Narratives

[T]he poetics of the open work is peculiarly relevant: it posits the work of art stripped of necessary and foreseeable conclusions, works in which the performer's freedom functions as part of the discontinuity. . . . Every performance *explains* the composition but does not *exhaust* it. Every performance makes the work an actuality, but is itself only complementary to all possible other performances of the work. In short, we can say that every performance offers us a complete and satisfying version of the work, but at the same time makes it incomplete for us, because it cannot simultaneously give all the other artistic solutions which the work may admit.
—Umberto Eco, *The Open Work* (1962)

Inspired by the appearance of what he perceived to be a notable shift in aesthetics across an entire spectrum of art, informing the works of artists from Jean Dubuffet and Pierre Boulez to James Joyce, Umberto Eco in *The Open Work* explores the radical differences in the aesthetics informing traditional and modern art. Like the hypertext narratives "Forking Paths," *afternoon*, and *WOE*, the works of modernists such as Henri Posseur, Alexander Calder, and Mallarmé leave their sequence or arrangement either to chance or to their audiences, providing them with a multiplicity of possible versions in which they can be experienced. Whereas traditional works appear to possess singular, determinate meanings, these modern "works in motion" seem constructed to provide their audiences with "a field of possibilities . . . a configuration of possible events, a complete dynamism of structure . . . and a corresponding devolution of intellectual authority to personal decision, choice, and social context."[34]

My version of the structure of *WOE* as a network of snippets of personal history, a chronicle of creation, and an invented narrative is thus both reinforced and modified by my knowledge of Eco's aesthetics of the open work. On one hand, my awareness of this aesthetic prompts me to see *WOE* as the paradigm of the open work, one that can embrace divisions normally insuperable in print narratives: commentary on the act of creation, the mechanics of production, the convergence of voices, past and present, the snatches of experience that become the grain that irritates, the core we pearl over to become the stuff of fiction. *The Open Work*, however, also provides me with a schema for recognizing the discontinuities in *WOE* as endemic to the open work, its indeterminacies the source of the narrative's rich field of possibilities. I have, in a sense, a metascript that also enables me to be comfortable with the very inconclusiveness of my reading, with its inability to account for everything I have discovered in *WOE*.

Even in interactive narratives, where we as readers never encounter anything quite so definitive as the words *The End*, or the last page of a story or novel, our experience of the text is not only guided but enabled by our sense of the "ending" awaiting us. We truly do read, as Brooks argues, in "anticipation of retrospection."[35] Our predictions enable us to minimize ambiguities, and to perceive words in an already largely determinate context, even when we move through a text knowing that the very words we read can and may crop up in entirely different contexts. The anticipation of endings is, in this sense, integral to the act of reading—even when no tangible, final ending exists. Ultimately, we cannot separate the desire for an ending—which might resemble Conrad's longing or Benjamin's sanction in the epigraphs beginning this chapter—with our need to create contexts for the perception of what we read in the immediate sense by anticipating what may follow in the future. When we read, prediction enables us to create contexts for words and phrases that guide our interpretation of their meaning in an action that appears to unfold simultaneously, not in discrete stages in time.

So when we navigate through hypertext fiction, we are pursuing the same sorts of goals as we do when we read *Our Mutual Friend* or *Love Story*—even when we know that the text will not bestow upon us the final sanction of a singular ending that either authorizes or invalidates our interpretations of the text. Because our sense of an "ending" does not derive explicitly from the text itself in the case of hypertext fiction such as *afternoon* and *WOE*, reading these narratives sheds light on what, other than the physical ending of a story, satisfies our need for endings or closure. We rely on a sense of the text as a physical entity in reading both interactive and print narratives, on a sense of having finished reading all of the book's pages or having visited most of a narrative's places, of having grasped the spatial form of *Mrs Dalloway* or *The Good Soldier*, of having arrived at a space that does not default in *afternoon*, or of having incorporated the contents of the periphery with the hub in *WOE*. Our sense of arriving at closure is satisfied when we manage to resolve narrative tensions and to minimize ambiguities, to explain puzzles, and to incorporate as many of the narrative elements as possible into a coherent pattern—preferably one for which we have a schema gleaned from either life experience or from encounters with other narratives. Unlike most print narratives, however, hypertext fiction invites us to return to it again and again, its openness and indeterminacy making our sense of closure simply one "ending" among many possible. It is often impossible to distinguish between explaining a work and exhausting its possibilities in the sense of the ending we experience when we finish reading *The Good Soldier.* My readings of *afternoon* and *WOE*, however, explain the versions of the texts I have experienced as I navigate through the hypertext without exhausting the number of other possible versions and explanations I might experience on other readings. If we as readers truly do long for a sense of an ending as the starving long for loaves and fishes, it is not the definitive, deathlike ending foreseen by Benjamin: a plausible version or versions of the story among its multitudinous possibilities will suffice equally well.

The Intentional Network

> [Open works] are, therefore, still a form of communication, a
> passage from intention to reception. And even if the reception
> is left open—because the intention itself was open, aiming at
> plural communication—it is nevertheless the end of an act of
> communication which, like every act of information, depends
> on the disposition and the organization of a certain form.
> —Umberto Eco, *The Open Work* (1962)

If we follow the thinking of theorists like Wolfgang Iser, aren't all
works more or less open, in that they invite readers to complete them,
to breathe life into the cluster of signs on the page that make up characters like Raskolnikov? Not really, Eco argues, since Dostoyevsky,
like the creators of most conventional works, has merely arranged a
"sequence of communicative effects" so that each reader "can refash-
ion the original composition devised by the author."[1] No matter how
much hatred I donate to Raskolnikov, how many memories of corrupt
cops and lousy officials I pump into my interpretation of him, I am
more or less falling into the lockstep Dostoyevsky has envisioned for
me. All my flights of imagination, all the remembered slights I dredge
up from my past to animate minute ciphers on the page go into
fulfilling a scheme already envisioned by a now-dead author. My belief
that I am creating something anew when I read *Crime and Punish-
ment* is as wrongheaded as the protagonist in E. T. A. Hoffman's "The
Sandman," who mistakes an automaton for a living woman, believing
that motion and animation are tantamount to life. I may be performing a fair number of operations as I read, as we have already seen, but
I am not, when I read conventional texts, so much bringing into being
a new version of the text as I am running through the paces dictated by
its blueprint.

However, the creators of modernist works such as *Finnegans
Wake* or Boulez's Third Sonata for Piano, Eco notes, did not construct

determinate sequences that would prompt readers or auditors to search for singular, definite meanings in them. To interact with these works, you simply plant your feet in the thick of what Eco sees as "an inexhaustible network of relationships," taking advantage of your freedom to choose points of reference, of entry, and of exit into the text.[2] When I dip into *Finnegans Wake* and tackle its layers of allusion, I fully realize that I might very well interpret the same passages in an entirely different way next Tuesday or next year. The text itself is littered with neologisms forged from as many as ten different etymological roots, each touching on a network of often highly divergent submeanings, which, in turn, allude to other words and other submeanings scattered throughout the book. Depending upon which meanings and submeanings I choose to focus on, the content of *Finnegans Wake* can appear to change each time I delve into it. My interpretation is merely one among many possible performances of its multifold meanings. Yet the book itself is still the same; what changes is my focus.

Compare this, however, with the eponymous book in the Borges short story "The Book of Sand." Swapped for a rare first edition of the English Bible, the Book of Sand is a nightmare of infinitude—it never offers the same page to any reader more than once. Not only do the words and pages themselves change, the hapless narrator discovers, as he opens, closes, and riffles through the book, but the very folio numbers change. No matter how many times he turns the pages, he recognizes nothing. No matter how far forward or how far back he pages, he can never quite grasp the pages that begin or end the thing between his trembling fingers. Finally, feverish from weeks of sleepless nights spent attempting to chart the book's limits, the narrator, hoping to be rid of the infernal book forever, sneaks it in among the nine hundred thousand dusty volumes of the Argentine National Library. Ever changing and physically inexhaustible, Borges's fictional Book of Sand represents the quintessential open work. Each part of the Book of Sand serves to reorient its readers as it reveals yet another virgin page for examination, the appearance of each fresh page altering their conceptions of the whole, and, consequently, of the role played by the contents of the page in the entire schema.

On the continuum between open print narratives like *Finnegans Wake* and the fictive, limitless Book of Sand, hypertext narratives float somewhere between the limits: physically multiple, unlike Joyce's novel, yet also physically limited, unlike Borges's fictitious book. As we have seen, the space lurking between segments of text— which requires readers to project links and connections into them—

the heightened indeterminacy of hypertexts, and their lack of singular, determinate closure make the likes of *afternoon* and *WOE* much more open than many of their modernist counterparts. On the other hand, when I page through *Ulysses,* I do not need to arrive at an interpretation of James Joyce's work to keep reading. Even if I haven't a clue what is going on or what on earth I have been reading, I can still push valiantly on ahead—unlike the readers of "Forking Paths," who, because they couldn't quite latch onto the key words Moulthrop had used to link segments, were stalled, unable to continue their reading. Their ability to make any headway into Moulthrop's text stemmed from their discovery of a set of navigation commands the author himself had forgotten existed, tools that enabled them to skip outside the network of scripted encounters with the text Moulthrop had envisioned. In one sense, "Forking Paths" is an open work in that the hypertext has nothing remotely resembling a set of "necessary or foreseeable conclusions," and its structure of the text is certainly, even literally, dynamic—both integral to Eco's definition of openness. No matter how many times I open *Finnegans Wake,* page 16 will not change, but each time I read through *afternoon,* the sixteenth place I encounter, not to mention my entire reading, can be different.

Yet, neither "Forking Paths" nor its more accessible counterparts *afternoon* and *WOE* fulfills Eco's vision of a text that represents the "devolution of intellectual authority to personal decision, choice, and social context."[3] As a reader of *afternoon* or *WOE,* I can end up at the mercy of the tangled skein of hot words and invisible defaults—where even my choice of paths can be contingent on just how well I have fulfilled Joyce's blueprint for realizing his text. Unless the author chooses to flag the hot words in each place so that I know these are gateways to other, related spots in the hypertext, I may not even know the difference between when I have moved by choosing the "right" words and when I have simply moved along a default path (which I can determine only by backtracking and checking out what the default option looks like). Defaults, usually the most straightforward connections between segments, can also be the most slippery, the most elusive, since they are invisible and can also remain inert and resist me until I complete all the right moves, choose all the right words, trace the proper pathway through the network.

Of course, we are also dealing with a fledgling, perhaps even embryonic, form of an emerging technology, a medium lacking stable conventions to curb its creators and guide its consumers. Like so many extensions of our sensibilities, hypertext is an odd creature, one that simultaneously promises more autonomy for its readers while offering authors a degree of control unthinkable with more conventional materials—seemingly as two-faced as any politician's promises. For example, Storyspace, the same hypertext software that provides writers with the capacity to attach guardfield conditions to segments of text, enabling them to dictate more or less fixed orders in which texts can be read, also features three interfaces offering readers incredibly varied degrees of autonomy. One restricts readers to navigating via default connections or hot words (see fig. 3). A second lets readers choose between defaults, paths, or hot words (see fig. 4). The third (see fig. 5), complete with a cognitive map, frees its readers to do all of the above—as well as to wander blithely through the hypertext via the map, completely disregarding every connection its author has so painstakingly crafted.

There is, of course, a problem with choosing this last and most autonomous route through the text: reading a narrative in a nearly random order can considerably narrow the distinction between fiction and life. Whereas fiction pleases us with its consonances, its patterns and gestalts, its symmetry and predictability, life can be chaotic and unpredictable, all sense of orderliness or pattern possible only at the distance conferred by retrospection after the passage of years. To encounter fiction outside any established order is to enjoy a dubious bit of freedom, less like an aesthetic experience and more like dicing with life itself. It can prove challenging, frustrating, puzzling, even occasionally utterly defeating—as the readers of "Forking Paths" discovered when they glided straight through what should have been points of closure and found a character who had just taken a bullet between the eyes conversing quite comprehensibly in the very "next" segment of text. Order gives us some of the delight we take in fiction, the comforting sense that things are predictable, stable, and knowable, that effects always have causes that can be traced, and causes effects that can be discovered, the sense that everyone murdered, mugged, arrested, or convicted merely receives what he or she has coming to them. While it is theoretically possible to create a text that could be read more or less randomly, most likely the readers who could take pleasure in it would need to have evolved a set of entirely different aesthetic expectations, satisfactions, and objectives than those of us accustomed to print and its literary conventions currently possess. Ultimately, the Book of Sand, that unfathomable treasure for which Borges's narrator swaps a rare, black-letter Wiclif Bible, becomes a horror, a nightmare of infinitude that seems to exceed even the boundaries of life itself.

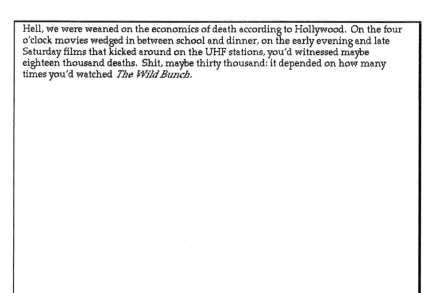

Hell, we were weaned on the economics of death according to Hollywood. On the four o'clock movies wedged in between school and dinner, on the early evening and late Saturday films that kicked around on the UHF stations, you'd witnessed maybe eighteen thousand deaths. Shit, maybe thirty thousand: it depended on how many times you'd watched *The Wild Bunch*.

Fig. 3. If you rely on the Storyspace Easy Reader, you can navigate through a hypertext like "I Have Said Nothing" via only default connections and linked words.

Conventional print fiction can seem everything life is not: it is fixed, it has a definite (and finite) form, and it resists the ravages of time far better than most of us ever will. Entering these miniature constructed worlds can seem like assembling a jigsaw puzzle. No matter how creatively you juggle the pieces, you are obliged to put the thing back together exactly as its creators intended, the same way Rhett Butler inevitably walks out on Scarlett O'Hara in *Gone with the Wind* no matter how you had wanted the story to turn out. Even when it is neither directly retrievable or verifiable, authorial intention, at a certain level, is inescapable. That intention, nonetheless, cannot dictate our every movement through any text, not simply because language is hopelessly slippery and fundamentally indeterminate but because, as readers, we inevitably bring a vast arsenal of tools to bear on anything we read—a world bulging with literary conventions, modern novels, B movies, snippets of psychology, plus all the wisdom and knotted scars of lived experience. And, perhaps more important, the majority of that experience is shared with the writers we read.

But still you knew

But still, you knew, as every kid did, that there was a definite, comforting order at work somewhere behind the scenes. The people you knew well, the ones whose names headed the rolling credits, the ones who had weeks named in their honor on the late-night movie slot, the ones who were important, they stuck around for the action.

The ones you barely knew or the ones you disliked—the spurned suitors and jealous deputies and cattle thieves, the Elisha Cooks and the ones without names, with forgettable faces or lousy teeth—they were thrown to death as a sop, something to stop the hunger pangs until the main course rolled around.

There was a neatness to death there, an admirable economy to it. Principals lingered like the sun at a summer evening's end. Character and supporting actors fell like dusk around the winter solstice, their lives running out in a shot or two. Extras snuffed it, their goings as brief and unremarkable as fruit flies'. They dropped two or three to a frame, and we thought nothing of it. It was natural; it was expected; it came with the territory.

Fig. 4. Readers relying on the Page Reader may move through hypertexts by defaults and linked words or browse the paths branching out from each segment of text.

Schooled in the same literary conventions that authors either obey or violate, it is a relative rarity for us to brush up against anything remotely unexpected, let alone startling or disorienting. Even if Ford intended us to construct a vast and intricate set of chronologies in our heads as we wend our way through *The Good Soldier*, his grand scheme does not condemn me to tracing maps of the novel's layers of time with every paragraph that I consume. I can just go ahead and read the book in a highly conventional way, since Ford has provided an abundance of clear-cut cues that tell me exactly when and where each segment of text falls in both narrative- and story-time. Likewise, I can read *The Alexandria Quartet* without picturing the narrative as a vast spiral that tunnels through representations of reality from the most superficial to the most fully informed. Understanding that the four

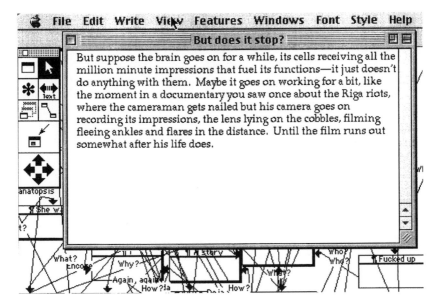

Fig. 5. With the Storyspace Reader, you can also use one further means of navigation: the cognitive map enables you to explore the hypertext independent of any links or connections its author may have forged.

books conform to this overall structure might help me interpret more quickly and easily the events I read, but since the tetraology adheres reasonably closely to literary conventions, even a far from ideal reader who has hitherto neither heard of nor experienced spatial form could probably comprehend it just as well.

In many respects, the book represents a highly sophisticated use of a relatively primitive tool, the printed word, where every piece of information, regardless of weight, nuance, and complexity, is relegated to the same physical level, our only alternative to linearity the footnote or endnote clumsily worked into the margins. The physical contents of any book are the same for an expert as for a neophyte—one might sweep over the same ground more rapidly and efficiently than the other, but the sheer number of words to be skimmed, digested, and, perhaps, assimilated remains constant regardless of the identity of the reader. And, spatial form notwithstanding, the machinations authors wreak on novels, the hoops they oblige their readers to prance through, are all right there, naked, in the text for anyone to see, even when the intentions that may have prompted Cortázar to set up *Hopscotch* in the order he has are veiled and irretrievable. There is no hidden technology that makes the pages turn, no need for me to mull over the process Cortázar or Borges or Ford might have used to spin their stories, and the setting of type—hot or cool—has nothing to do with the reading, meaning, or experience of the book.

On the other hand, hypertext fiction, we should never forget, is digital. There are always at least two other texts lurking beneath whatever we read, both of them like palimpsests, medieval parchment painted over for reuse where the topmost text was superimposed on whatever had existed beneath it. Like a palimpsest, most of these other texts are secret, hidden, even invisible. Most of us never glimpse the code that enables hypertext software to run—an environment that itself already determines how authoritarian, how automatous the fiction may seem to its readers—just as few of us can do more than guess at the scripts hypertext authors generate in the creation of a network that will act as choreographer to its readers' realizations of the text. We might be able to guess, as the readers of "Forking Paths" did, that the hypertext has used hot words to connect segments of texts once we have tried every other navigation strategy and come up empty-handed. But I cannot know for certain that I have hit on the "right" words when I can also move by default, as readers can in *afternoon*—particularly since I know that authors can attach guardfields enabling the text to change, even to generate a different default if I back up to verify why I have moved from one place to another.

To make my guesses about the author's script still more difficult, there are no fewer than three different kinds of possible hypertext connections—defaults, paths, and links using hot words or images—which are inherently different, although all can appear on a map as ties between segments. The distinction between navigating by default and navigating by hot words is a little like the distance between channel surfing your way through cable TV and surfing your way through the Internet: one takes a good deal more interaction, dexterity, and thoughtfulness than the other. If I spend, say, five minutes carefully scrutinizing the text of a segment in *afternoon*, trying to figure out what might fit Joyce's concept of "words with texture" to trigger a hot-word link, I might not know whether I have moved because I have succeeded in matching my preferences to Joyce's particulars or because

Joyce designed the text so I could move from one place to another by default. Even choices that appear on the Path menu might appear in one context but not in another, according to whether I have or have not met certain conditions. Hypertext is language on top of author-generated scripts on top of codes written by programmers. Although the scripting may be as artful as the prose—and its creation more arduous and time-consuming than the writing of *Moby-Dick*—these are entirely invisible strata of text, levels that manipulate us but, if the author chooses, we can never see.

Reading, Navigation, and Intention

> It is only because an artifact works that we infer the intention of an artificer. . . . Poetry succeeds because all or most of what is said or implied is relevant. . . . In this respect poetry differs from practical messages, which are successful if and only if we correctly infer the intention.
> —W. K. Wimsatt and Monroe C. Beardsley, "The Intentional Fallacy" (1954)

> If [the reader's] claim to validity is to hold, he must be willing to measure his interpretation against a genuinely discriminating norm, and the only compelling normative principle that has ever been brought forward is the old-fashioned ideal of rightly understanding what the author meant.
> —E. D. Hirsch Jr., *Validity in Interpretation* (1967)

To utter the word "intention" today in conjunction with authors, meaning, or interpretation is to invite anything from a sneer to out-and-out castigation in academic circles. With the publication of "The Intentional Fallacy," Wimsatt and Beardsley followed up on the New Critical separation of author and text by arguing persuasively that a grasp of authorial intention was irrelevant to the interpretation of texts. The poem either works or it does not, the argument runs, and if the poem is successful, authors' intentions are embodied in their works—we have no need of extratextual information about what they had in mind in the design of their texts.[4]

With the appearance of psychoanalytic criticism, such as Lionel Trilling's "Freud and Literature," the concept of intention as a critical category was tarnished still further, as Trilling and others like him argued that intentions are usually unconscious, generally irretrievable, often completely unknown to even authors themselves, and relatively useless when it comes to finally determining the meaning of a work.[5] Arguably, the reappearance of intention as the much-valued keystone of E. D. Hirsch's solution to indeterminacy in interpretation was the final nail in the coffin. In treatises such as *Validity in Interpretation*, Hirsch has repeatedly argued for the central role the category of intention plays not in enabling interpretation but in anchoring it, stabilizing it, and removing it entirely from the hurly-burly of indeterminacy, intertextuality, and what for Hirsch can only be the horror of postmodern claims heralding the Death of the Author. By insisting that the author's original, intended meaning represents the most valid interpretation of any text, he shores up the text amid shifting postmodern sands and makes its meaning both determinate and singular. Where the category of original intent can still be seen as central to enabling interpretation—as in the rules of baseball and the United States Constitution—in the hands of both Hirsch and lawmakers it is patently used as a tool to defend and sanctify certain interpretations and to limit others, a Fort Apache throwing up walls against the slings and arrows of indeterminacy.

In any case, few theorists or readers would care to claim that we need concern ourselves with authorial intention when we read: the narrative either works or it does not, and if it works, everything we need is in the text. But this is accurate only when we refer to print narratives. After all, in order to continue reading, I do not need to decide whether Iris Murdoch's *The Unicorn* is a Gothic novel, an extended conceit about the relationship between authors, readers, and texts, or a fable about Christ-figures and scapegoats. I can simply continue reading, oscillate between interpretations, waver between all three— or lose myself for the space of a few hours in what seems like a good Gothic revel, thick with wrecked romances, drugged heroines, and even contemporary damsels in distress, without knowing that any other ways of reading the novel exist. Of course, I need to arrive at some conclusion as to what the narrative is about, or else I will not be able to even unconsciously form predictions as I read. In fact, I will not, by definition, be able to read at all. But I do not need to satisfy some unseen author's conditions about what my predictions should— or even must—look like, unlike the frustrated readers struggling in their various ways through "Forking Paths."

In hypertext fiction, unlike its print counterpart, authorial intention is palpable. Although my intentions in writing the words you read might put a certain topspin on them, might try to persuade you to assume my viewpoint on hypertext fiction, you do not need to intuit the shape of my intentions, let alone share them, to make sense of what I read. But if you were to read my hypertext fiction "I Have Said Nothing," the satisfaction you might derive from reading it is contingent to a large extent on your ability to replicate a certain approach to the text, to reach certain conclusions about the relationship between the stories it relates and the ties that bind them. The very cues you need to proceed from one segment to another have nothing to do with the words of the text you read and everything to do with the script of that other text, the author's script, the one you can intuit and certainly feel but never quite see. As you read *afternoon* or Stuart Moulthrop's *Victory Garden*, for example, you might scrutinize the names of places and paths, ponder the patterns behind the default connections or the logic that unites certain hot words and the destinations to which they are linked. You might mull over the relationship between the name of a path and the title of a destination, or the way a cognitive map might provide some hint about the shape of the narrative. It is not a case of your being a terribly analytical reader or one unduly concerned with how hypertext narratives work at some deep level. It is a matter, more often than not, of your being able to evolve a strategy for understanding and manipulating (or, probably more accurately, being satisfactorily manipulated by) a palpable, omnipresent subtext that is actually a distinct, discrete text in itself. If I read this authorial text incorrectly, refuse to deal with the intentions it represents, the hypertext can freeze me out, petrify my reading, imprison me within a single segment of text until I behave, satisfy all the right conditions. Perhaps Barthes—who met his untimely end before the advent of HyperCard and the World Wide Web bestowed on *hypertext* the status of popular buzzword—was a tad premature: the Author, it would seem, is far from dead. Within hypertexts, he or she can be kept more omnipotent and omnipresent than in print (and perhaps even in life), embodied as ghosts in the machine, authors of what I will call the "intentional network."

In one sense, by introducing virtual layers into texts, hypertexts leave spaces into which authors can insinuate their expectations for the variety of ways readers might interact with their trace, the text they leave behind. In print fiction a text is all surface. Intention there can be visibly embodied in all the puns, twists, and spins an author can wreak on literary conventions, like Sterne's blank pages, Cortázar's two different versions of *Hopscotch,* and the mutually exclusive versions of the eponymous babysitter's evening in Coover's short story. In hypertext fiction, the author both tells a story and designs an experience that unfolds in time—not the fixed and immutable narrative a writer might create in print, but a series of potential interactions that span both time and space. The intentional network—all the structures in the hypertext that either aid or restrict my navigating through it—shapes my experience of not only *how* I read but also *what* I read: providing me with paths to follow or words to choose, enabling me to view certain choices and not others. If I am in hot pursuit of the answer to a question, confirmation of a hunch, or the opportunity to end my reading, I need to be at least as concerned with interpreting the structural details and nuances of the hypertext as I do its content. Even when I am reading casually, the intentional network—made up of guardfields and defaults, link labels or icons and window titles, hot words and cognitive maps—shapes the options I can choose and the trajectory of my reading. In any case, my awareness that they exist, that they have been designed by an author and integrated into the text I read makes it difficult for me to see these features as insignificant, even if I could navigate through a hypertext easily without paying any attention to them:

> It must be emphasized that intention alone is enough to give noise the value of a signal: a frame suffices to turn a piece of sackcloth into an artifact. This intention can, of course, assume all sorts of different forms: our present task is to consider how persuasive they must be in order to give a direction to the freedom of the viewer.[6]

The structures in the intentional network can operate nearly invisibly, seeming neither to conspicuously guide my reading nor to frustrate it. In these instances, where the design of its structure seems so intuitive, the content so self-evident, I need only wonder about the text as an intentional object when I seek a fresh reading of it, or when I feel puzzled by, say, a character's reasons for revealing to us what she does. But "persuasive," in the context of hypertext narratives, does not seem entirely accurate; *informative* would, perhaps, better describe the accouterments that make up the intentional network. In order to provide readers with a sense of what Eco sees as an oxymoronic "directed freedom," authors of hypertext narratives may use names, recurrent words, or phrases, puns, and allusions to draw attention to the frames through which readers might fruitfully encounter their work—providing information sufficient to enable readers to realize its possibilities, bridge its indeterminacies, and, of course, return for another stab at the whole process.

Of course, it is hardly surprising that we should also encounter insistent and even virulent versions of the intentional network. All hypertext narratives currently in circulation (and probably every one that will be written well into the twenty-first century) were created far beyond the boundaries of literary conventions, the very signs that enable us to make sense of *In the Labyrinth* as well as this month's *Reader's Digest.* A hefty part of the lure for writers working with hypertext lies in its invitation to what is, in a sense, lawlessness, the freedom to evade all the usual rules, as well as to revel in the glorious freedom to invent, to endow things like maps and paths with the power to signify. On the other hand, convention is, as we have seen, a great enabler: it provides us with schemata that enable us to perceive what would otherwise remain insignificant and, thus, all but invisible. Convention, or at least its long shadow, is also inescapable. We cannot help but lean on conventions and all the old, familiar schemata regardless of how alien, how high-tech the environment—witness the store of literary expectations, readerly strategies, recollections of canonized texts and pulpy novels, as well as psychological rationale gleaned from movies, brought to bear on my interpretations of *afternoon.* Hypertext has, ineluctably, inherited a number of print conventions that have trailed into the medium on the lengthy coattails of print literature. The difficulty—and what makes the intentional network so obtrusive at the moment—is that literary conventions have, not surprisingly, been hurled willy-nilly into the breach, used tactically where we are accustomed to dealing with them strategically in print. Evolved over centuries and hordes of works, print literary conventions have long been part of a code shared by authors and readers, a set of writerly rules and readerly expectations that meant that even wildly inventive texts like *Ulysses* could be understood, absorbed, and even enjoyed by readers. When hypertext authors import conventions from the world of print, however, they have used them to explore and evolve an aesthetics of hypertext fiction. As a result, every effect aimed at is strictly experimental, and literary conventions are wielded in whatever way suits local tactics, so that, for example, the role or significance of default connections can vary between authors, between works by an author, even wildly within a single text. Sometimes, as we saw in *WOE*, this lends itself to a particularly rich playing off of expectations you never knew you had, as when I discover the same "he" sleeping with two entirely different "she"'s in two continuous places joined by a single default.

Of course, I can detect the logic ticking away behind some of the elements of the intentional network. In *afternoon,* for example, a sequence of places narrated by Peter's ex-wife, Lisa, is connected by a path called "Hidden Wren," an image Lisa includes in her musings:

> I do know what you feel. You make some choices, you begin to see a pattern emerging, you want to give yourself to believing despite the machine. You think you've found something. (It's a beautiful image, really, the hidden wren—I told you I thought he was a genius . . .) I think he means it to be the clitoris, all nervous and yet somehow self-contained—a bird's perfect, really (although I'm being too literal I suppose, it's all images, isn't it?) That's why I'm sorry I have to end it for you so soon.[7]

When I locate the path name "Hidden Wren," in the Path menu, I know that choosing it will route me into the strand involving Lisa's voice—a string of places ending with what may be her suicide ("I'm sorry I have to end it for you so soon") or perhaps simply her exit from a brief, puckish, and definitely uninvited intrusion into what she sees as Peter's narrative structure. When I select the words "hidden wren" in the segment of text cited above, though, believing them to fit Joyce's description of "words with texture," I do not move further along the "Hidden Wren" sequence. It is only when I hit on choosing just "wren" that I move directly into the next segment of the sequence (which I can verify by backing up, or by fudging a bit and using the Storyspace program to crank open *afternoon* and check out its guardfields). As anyone with the slenderest experience of computers doubtless knows, even moderately sophisticated programs can seem to manipulate information less adroitly than five-year-olds who can match a word or phrase to one that more or less resembles it. Many applications can only distinguish the difference between a direct hit— the selection of "wren"—and a miss, making my choice of "hidden wren" a miss. Although I believe both "wren" and "hidden wren" should trigger a link and probably Joyce would have agreed had he been peering over my shoulder, the logic governing guardfields and their Boolean strings cannot match my near miss with its specified hit—a situation that will probably be rectified with the introduction of more sophisticated devices like fuzzy logic that can identify similarities between words.

More often than not, hypertext narratives defy rather than correspond to our expectations, staking out our trails through the text not with helpful bread crumbs or bowed branches but with dense thickets of puns and allusions, jokes and rapid reversals of expectation. In *Victory Garden,* an accessible congregation of narratives about the lives of friends that come together during the Gulf War, an intricate network of associations, allusions, and submeanings becomes itself a palpable part of the reading, a rich text brimming with double entendres, timely cracks, and puns. In one segment, the link word "you" in the text propels us straight to the place "Dear You," and a text that includes the contents of a letter addressed to a character named Urquhart, who goes by the nickname "U." Elsewhere, choosing the link words "what we don't see" directs me to a segment where one of the characters juggles her attention between flickering images on the TV in front of her and snippets of conversation with whoever is sitting alongside her—only we, the readers, do not "see" either: neither the television program nor the other body shunting words back and forth is identified. In the same vein, when I choose "male faces" in the text of the place "Where Are You," I arrive at the beginning of a narrative strand about the narrative's ardent feminist-activist as she battles the patriarchy—not exactly what you would have expected to find at the end of that particular link.

Sometimes, almost perversely, the connections in *Victory Garden* eschew associations or allusions, as if trying to keep me on my toes, to caution me not to lean too heavily on my expectations. When I embark on the path called "Memories," I expect the word "remember" in the text of one place on the path to keep me steaming ahead through "Memories," since the verb and the noun are related by more or less the same etymological root. For all this, the word "remember" has nothing to do with the path "Memories," leaving me to cast around for other ways of sticking to the same path. Even the default connections here frequently upset my readerly expectations, providing nothing resembling an orderly, accessible, single path through the narrative—not coincidentally, exactly the absence of a master narrative, the no-default condition, central to the definition of interactivity in chapter 3. Instead, the logic behind the defaults keeps changing, chameleon-like, as I plow ahead, forcing me to keep revising my predictions about where the next default may take me. In the midst of a strand about the Runebird family—two college-age women, one stationed on the Desert Storm front in Saudi Arabia—I discover that I have moved by default to a place called "Bird-Fiver-Two." Because the context of the preceding place concerns the history of Emily Runebird, I assume that the default will transport me to Emily and the throes of Desert Storm, since Emily's family name contains the name "Bird" and the B-52 (itself a kind of "bird") was the Air Force workhorse ply-ing the skies above Iraq. Instead the contents of "Bird-Fiver-Two" are merely a paean of sorts by a nameless first-person narrator on the sturdiness of the B-52 construction. While a "You" may sometimes yield a "U," occasionally in *Victory Garden* a "Bird-Fiver-Two" is just a Bird-Fiver-Two.

Just How Open Is the Open Work?

Certainly, this palpable and often intricate network of jokes, puns, and allusions seems a far cry from Eco's vision of the complete "devolution of intellectual authority to personal decision, choice, and social context."[8] Or is it? Near the beginning of *The Open Work*, Eco meditates on the nature of what he terms "the work in movement":

> We can say that the "work in movement" is a possibility of numerous different personal interventions, but it is not an amorphous invitation to indiscriminate participation. The invitation offers the performer *the opportunity for an oriented insertion into something which always remains the world intended by the author.*
>
> In other words, the author offers the interpreter, the performer, the addressee a work *to be completed.* He does not know the exact fashion in which his work will be concluded, but he is aware that once completed, the work in question will still be his own. It will not be a different work, and, at the end of the interpretative dialogue, it may have been assembled by an outside party in a particular way that he could not have foreseen. The author is the one who proposed a number of possibilities which had already been rationally organized, oriented and endowed with specifications for *proper* development. (19; emphasis added)

A last turn of the screw and we seem to have been cast back, once more, onto Plato's disparaging assessment of writing—with the text not only scurrying back to its parent but also always saying, fundamentally, the same old thing. You can fill in the details and emotions, even choose the order in which you prefer to realize the thing, but the text is ultimately someone else's commodity, and all the sweat, toil, and tears we loan the text are simply the lubricants that grease the wheels of an already complete mechanism. Ultimately, as far as Eco is concerned, the liberty we enjoy as we explore the open work is simply a set of oxymoronic, circumscribed freedoms the creator intended us to realize right from the outset, nothing more. Just when theory encourages us to believe the line between author and reader is never as definite or as impenetrable as the old Berlin Wall used to seem, the physical hallmarks of analog media intrude to reassert all the old distinctions—and we are left with the original triumvirate of author, text, and reader, ranked in order of descending importance. Our first impulse might be to ascribe Eco's still fairly traditional take on the open work to the fledgling state of hypertext systems and their ilk at the time he was busy putting the finishing touches on his theory. In 1962, when Eco published *The Open Work*, hypertext was, after all, just an abstract concept still being fleshed out by Douglas Engelbart in the paper he would publish later that year, "Augmenting Human Intellect: A Conceptual Framework."[9]

The experience of readers struggling with "Forking Paths," as well as my own experiences in reading and navigating through the secondary, intentional network of defaults, paths, and links, however, might also encourage us to look favorably on Eco's insistence that my experience of reading interactive narratives represents simply an opportunity for an oriented insertion into something that always remains the universe intended by the author. In this view, the author is not only *not* dead, the author is deathless as well as inescapable. The very act of reading is predicated on his or her existence, and in interactive narratives as in print stories, we can never truly banish this given from our reading experiences. So the interactive narrative remains the paradigm of Eco's open work because the work of art is still, as Eco notes elsewhere, "a form of communication, a passage from intention to reception. And even if the reception is left open—because the intention itself was open, aiming at plural communication—it is nevertheless the end of an act of communication which, like every act of information, depends on the disposition and the organization of a certain form."[10]

Even though we receive the work by completing it, even if the "words that yield" in *afternoon* have been chosen to invite us into the role of cocreator of the narrative, our readings still constitute the reception of an intention. *Afternoon*, after all, will not miraculously yield up links when I select words that Joyce has found uninteresting and has excluded from his link-default-path structure. At the same time, the hypertext may well resolve into unimagined combinations and sequences during any single reading, since hundreds and even thousands of possible versions of the text exist—many of which Joyce has, doubtless, never so much as envisioned.

That leaves us with a very real conundrum. In one sense, we are faced with a medium that promises to increase the dynamic nature of reading exponentially with texts that actually change from reading to reading, with a range of choices and reading decisions that seem to offer readers an autonomy undreamed of in their experiences of print narratives. Readers engaging hypertext narratives, as we have discovered, can evade authorial control, rely on maps to create metaphors of their reading experiences, and decide when and how their readings are truly "finished"—just when and where the narrative for them comes to a form of closure. In liberating all the richly associative links between segments of text by lifting them out of the linear and syllogistic order endemic to print narratives, hypertext narratives invite readers to bridge the gaps between textual spaces, to leap into the breach with the same perceptions of connectedness, the same predictions that characterize our dealings with both the world around us and with print narratives.

But at the same time, hypertext fiction also presents its readers with an intentional network, an additional discursive structure that thrusts itself between text and reader, obliging the reader to engage in an already scripted interaction foreseen by an author who has fictionalized an audience for it. If I supersede this network and navigate by the map or by relying on directional arrows, I run the risk of being overwhelmed by the resulting cognitive overload—and of finding the narrative largely incomprehensible. Of course, the topographic map of narratives such as *afternoon* and *WOE* is itself an intentional structure—a schematic representation illustrating relationships between places and the paths connecting them, all quantities envisioned by the author. Even when I gaze on places such as the "Relic" strand in *WOE* and make decisions for further exploration, my choices are, to a large extent, guided as much by the names of paths and places as they are by the visible shape of the places and paths in the hypertext.

Where, you are doubtless tempted to ask, is the radical reconfiguring of author and reader that hypertext has ostensibly promised since the days when the technology was little more than Vannevar Bush's sketches of microfilm and illuminated desktops? Where is the genre that promises, as I suggested in chapter 2, to alter the way we think about books as intellectual property, to blur the lines between author and reader? We need look no further than the same medium that has given us the intentional network. Like any hybrid, hypertext has some of the characteristics of one parent, print, and some of the other, digital technology. If, at the moment, its phenotype seems to have more of the look of print than it does of digital flexibility, we can put it down to our peering so closely at something barely in its infancy. For all this, there are already signs of the old order whirling around on its foundations. In Deena Larsen's *Marble Springs,* readers can traverse a map of a frontier town, explore the biographies of its inhabitants, read poems describing events or characters in the town's history, and, most radically, add to the whole thing. In fact, you can write in nearly any portion of *Marble Springs,* as well as dream up new characters and fresh scenes. "By its very nature, *Marble Springs* can never be complete," Larsen writes. "Characters lead to other characters; connections lead to more connections until memory breaks under the strain of the web. The characters and connections you create become intertwined within the web of *Marble Springs.*"

Unlike the comments that litter the margins of the books I own, my additions to Deena Larsen's hypertext are all but indistinguishable from her text: only the two of us will ever know where her text begins and mine ends. But other readers will not—the writing instructions in *Marble Springs* invite readers to send their additions and new versions of it to its publisher, where they can be added to the growing hypertext. This, moreover, is no rhetorical invitation: new versions of *Marble Springs* have already debuted, a text where reading is not merely tantamount to the creative effort of the author but is essentially the same animal.

With hypertext, the whole notion of the author-ized text is, as Jay Bolter warns us, a suspect commodity. While few of us could lay hands on the paper, printers, and bindings to produce alternative, subversive versions of, say, *The Closing of the American Mind* or *Cultural Literacy* (and it would, in any case, take a good bit of scheming to insinuate them onto library shelves or into book stores), many of us can find ways of imperceptibly altering digital information. After all, hypertext is native to an environment where an original is indistinguishable from a potentially infinite number of copies, and where an entire economy of exchanges, downloading, sampling, and altering has made reinforcing notions of copyright tantamount to holding back an unruly sea with a child's finger stuck into a crumbling dyke—a situation heightened almost immeasurably by the possibilities for uploading, downloading, and linking represented by the World Wide Web.

In *Writing Space: A Hypertext,* Bolter reminds us of the peculiarly dated and print-bound notions of both copyright and author-ized texts in a brilliant riff on the standard (and usually unread) copyright notice that prefaces the contents of every book. Since he has full control over the electronic rights to *Writing Space,* Bolter can invite us to do whatever we want with it (see fig. 6), including take the money and run.

The publisher of Bolter's book of 1991, *Writing Space,* however, still retains all rights to the print versions, so Bolter reminds us that the rights are ours to play with only as long as the text is merely virtual, or at any rate, encased in hard plastic and not laid out on paper sheets (fig. 7). Yet, as we keep clicking through this seemingly endless copyright notice, the contradictory images multiply, a little like Rita Hayworth multiplied in the funhouse mirror in *The Lady from Shanghai,* where you cannot be exactly certain which lady wielding the gun is the one capable of pumping you full of holes. Perhaps this third addition (fig. 8) to the copyright notice has been tacked on by somebody with a sense of fun. Maybe, even, the entire sequence of notices has been cooked up by some puckish reader who has merely preceded you, and the real notice should just read the way these warnings always read (fig. 9). If you make it this far and still know who the "author" is, perhaps you have confused the hypertext author with the print Author—or maybe it is because you bought your particular copy of *Writing Space* direct from the publisher and have the Visa debit slip to prove it.

Interactive Stories—Who Needs Them?

The tale goes that Benjamin Franklin and a friend saw the ascension of a balloon in Paris and the friend after the show remarked something like, "What good is that?" Franklin answered "And what good is a newborn babe?"
 —David Freeman Hawke, *Nuts and Bolts of the Past: A History of American Technology, 1776–1860* (1988)

Now we arrive at last at the big question, the one that has haunted many of us the way it did a member of the audience at a lecture I gave on hypertext who pursued me timidly into the street: "I was embarrassed to ask such a silly question in there," she said, catching up with me, "but I just can't help myself. Why would any of us want things to be interactive?" Of course, it is the question nearly all of us pose, sometimes scoffingly, when we hear about the latest innovation, some invention aimed at enabling us to accomplish things we never dreamed of doing in the first place. Probably no one conceived of a need for a device like the telephone when they could communicate perfectly well by an efficient postal system—which, in Victorian London, featured as many as six deliveries a day. Or who could imagine that television, with its initial minute, fuzzy, flickering picture could

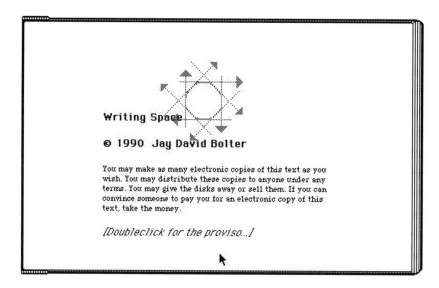

Fig. 6. The opening to the copyright sequence in Jay Bolter's *Writing Space: A Hypertext*. A twist on the familiar copyright, "look but don't touch" warning. The sky appears to be the limit, as far as what you are authorized to do with this text: if you can actually manage to sell electronic copies of this text—take the money and run.

one day temporarily flatten the booming film industry as quickly and efficiently as any blitz? Most of us tend to think of technology as merely satisfying pressing needs, helping us to delegate most of life's pure drudgery to machines, freeing our hands so we can use our heads more effectively.

Actually, however, technologies just as frequently foster new demands, stimulate new industries, provoke us to discover new ways of enjoying ancient arts. Consider, for example, the way in which the development of the motion picture, on the face of it, answered no apparent pressing need: for entertainment, people could always crank open a cheap novel, slip into a seat at the vaudeville, theater, or opera. Yet, even during the silent era, filmmakers like George Méliès were already experimenting with the special effects that were to become the staple of genres like horror and science fiction that, although popular in book form, had no close counterparts in the theater. And vast film sets, as well as heavy editing, enabled D. W. Griffith in *Intolerance* to tie together four stories separated by thousands of miles and

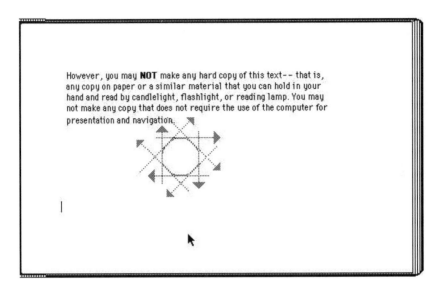

Fig. 7. Just when you think you know exactly what is permissible and what is verboten, the notice pops up again, telling you now just what you *can't* do with *Writing Space.*

hundreds of years against a background far beyond the realism possible on the stage—just as storytellers like Ford, Woolf, and Joyce can mimic the intricate tangles of memory, evoke the richness of associative thought, and remind us of the essential malleability of language in a way impossible with the spoken word alone.

Even in its present, primitive, ice box incarnation—the thing that keeps food cool but still requires daily deliveries of ice—hypertext and hypermedia encourage us to find new ways of satisfying the same old cravings that storytellers have both sharpened and sated for millennia. If part of the pleasure we take in stories lies in the ways they differ from life, in their orderliness, symmetry, even predictability, why should we be satisfied if they share one of their most important limitations—their singular endings—with life? Digital narrative mysteries could confer on us the chance to play out our ten or fifteen hunches and toy with contingency and probability in a safely enclosed little world where, one can imagine, you could utterly screw up, perhaps even get killed for your pains, and still bounce back on Tuesday to learn from your mistakes. Just as film genres were based on already existing types of novels and plays, we do not need to rack our brains to

As long as you keep the text in the electronic medium, you may also change it as you see fit and hand the changes on to others. You may want to indicate that you have changed the text. On the other hand, you may not, but then your readers will probably falsely assume that the original author was responsible for the text that you wrote. **All readers should be aware that anything in the text may have been added by someone other than the original author.** But, of course, this caveat applies in a Borgesian way to the previous sentence as well. Indeed perhaps this whole, crazy copyright notice was added by some mischievous reader. Perhaps the original author had a traditional copyright notice that read as follows:

[Doubleclick...]

Fig. 8. None of this exactly comfortably fits anyone's expectations of your standard copyright notice. Maybe it isn't—what guarantee do I have that some other reader hasn't tossed this entire sequence (or even just part of it) into Bolter's hypertext?

wonder what interactive narratives might look like as the technology grows more sophisticated. It is hardly difficult to imagine that writers working in print genres already based on predictions, contingencies, and premises—like mysteries, thrillers, horror, historical, and science fiction—would, in the future, want to capitalize on the mutability and multiplicity hypertext and hypermedia offer.

Digital narratives, as we will see in chapter 7, can marry fictive fantasies with graphic tangibility, as in a hypermedia narrative like *Who Killed Sam Rupert?* (1992), which equips you with a helpful and voluble assistant, a copy of the coroner's report, a crime scene, and a virtual stiff—and asks you to get to work on rounding up the culprit from a complex and colorful network of suspects. As Joyce and Moulthrop have already discovered, there are whole new aesthetic realms to be explored, mapped, and tamed—following on the first forays we have explored here: on the way *afternoon* explores the poetics of hypertext space, "Forking Paths," the pleasures of plots and their conclusions, and *Victory Garden*, the richness of a text made up of link words, place names, and default options.

At the same time, looking at the rudimentary workings of the

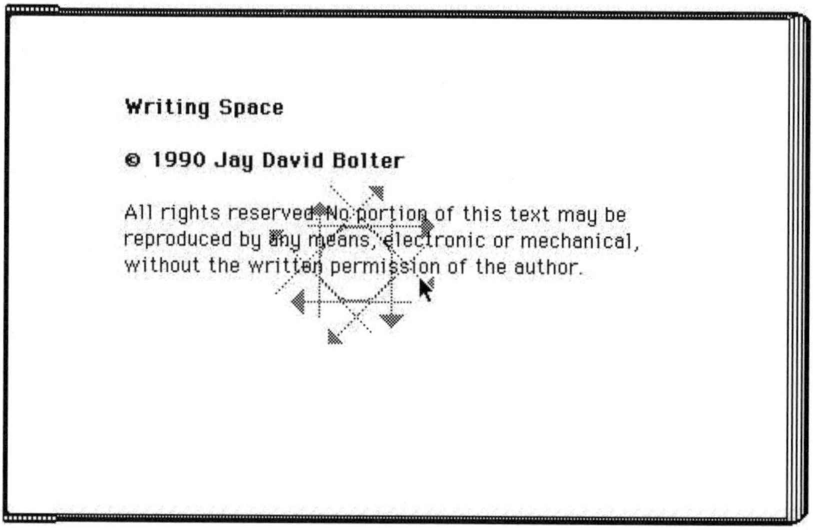

Fig. 9. What do you believe? Your Visa debit slip that proves you received this direct from the publisher? But some disgruntled hacker toiling away somewhere in the bowels of the company could have spiked Bolter's copyright message this way. So could another reader with a good sense of humor, contempt for copyright laws, and just about anything with plastic keys on it. Is this the true Death of the Author, or just the expiration of the Author-ized Text?

hypertext and hypermedia narratives currently in circulation, it is also difficult to predict what the medium, let alone its genres, might look like given a generation or two. Think of the set of wings the Wright brothers coaxed into flying at Kitty Hawk. That first airplane was able to take flight, all right, but you were not going to find anybody but a handful of reckless enthusiasts who could have foreseen its enormous utility, its vast role in shaping the world. You certainly could not have predicted the future of flight from the flimsy craft that barely made it off the ground, let alone envisioned jumbo jets trundling down runways in a hundred airports—much less an invention that could shrink the traversing of the globe from a feat that absorbed weeks and months into the length of a single day, giving us a world where a vacationing New Yorker can go sightseeing on Tierra del Fuego and Marshall Islanders jet to Las Vegas to play craps around the clock. We build technologies according to our desire to extend the repertoire of the human body and mind, and we can end up reconfiguring our lives and our ambitions around the new worlds they eventually open to us.

Technologies with vastly different capacities enable us to do radically different things with them. Yet, as we have seen, hypertext is not inherently democratic or liberating or egalitarian any more than it is implicitly more limiting, more authoritarian than print.[11] It can enable authors to create texts where readers experience information precisely as they want them to perceive it—no skipping ahead permitted, no skimming allowed, strictly no cheating possible. And yet the same technology also allows readers to potentially traipse all over a text, chop it to pieces, reassemble it, even circulate versions of the text that merge readers' contributions seamlessly, invisibly, undetectably, with the author-ized version. It is infinitely more open, more radical than print; it is unimaginably more controlling, potentially infinitely more restrictive than anything possible in print. Hypertext could lead to, say, an interactive textbook on English grammar that would not let you stray past a loathsome set of exercises until you managed to prove by getting every answer straight that you knew the subjunctive mood from the past imperfect tense. At the same time, the technology could just as easily lead to nonfiction texts where authors eschew championing a single point of view, and where, perhaps, collaboration between writers results in a reading experience akin to listening to jazz musicians jam, with each voice heard distinctly in a work of multiple facets and perspectives on a single subject. A hypertext like Christiane Paul's *Unreal City: A Hypertext Guide to T. S. Eliot's* The Waste Land or George Landow's *The Dickens Web* can have a paradoxical way of seeming more of a unity than a print collection of essays because the lines of thought, choices of subject, and points of view clearly and tangibly make contact throughout your experience of reading. At the same time, the voices within each place, even when segments follow one another on a designated pathway, can preserve their particular flavor, their tone, since each segment of text is distinct from those around it.

The characteristics of hypertext and hypermedia are, as yet, largely undefined, and, at any rate, predicting their potential from any current and primitive incarnations would be a feat on par with forecasting the emergence of fiction like "The Babysitter" after taking a peek at Gutenberg's modified wine-cum-printing press. At the moment, though, even in its rudimentary form, hypertext has the undeniable utility of providing us with a medium, a platform, that enables us to formulate questions we could not pose given our experiences with analog media. It may well represent a shift akin to that witnessed and bemoaned by Plato in his *Phaedrus*—a change that may alter not merely the way we represent our stories but also the stories we can tell. Like Plato, we can see enough of the new medium to realize that it has the potential to transform the way we think, what we say, even what counts as knowledge and information. Like him, we also know, essentially, mere glimmerings of the changes to come: it is a long road from *Phaedrus* to *Finnegans Wake.* It may well be a still further stretch from *afternoon* to the apotheosis of interactivity.

7

Millennium Stories:
Interactive Narratives and the New Realism

In contrast to our vast knowledge of how science and logical
reasoning proceeds, we know precious little in any formal
sense about how to make good stories.
—Jerome Bruner, *Actual Minds, Possible Worlds* (1986)

"[S]urrender . . . and the intimacy to be had in allowing a beloved
author's voice into the sanctums of our minds, are what the common
reader craves," writes Laura Miller.[1] Sven Birkerts sounds a similar
note of surrender in *The Gutenberg Elegies:* "This 'domination by the
author' has been, at least until now, the point of writing and reading.
The author masters the resources of language to create a vision that
will engage and in some way overpower the reader; the reader goes to
work to be subjected to the creative will of another."[2] As we saw in
chapter 1, like many readers with a slender experience of the medium,
both critics assume that, if interactive narratives do not spell the
Death of the Author that Roland Barthes described in his famous essay
of that name, interactivity will diminish the author's role, make it
nearly irrelevant—a fear, as we discovered with the intentional network, that is as lacking in substance as it is naive.

Strikingly, both Miller and Birkerts assume they speak for the
desires and predilections of the Reader, as if the *New York Times* best-
seller list or most of the books toted home by shoppers at Barnes and
Noble represent the Great Works of the century, the titles that find
their way onto college curricula and not ephemera that frequently go
out of print and are forgotten by the decade's end. A peculiar note of
triumph in her tone, Miller notes that the only people who seem to be
buying hypertext fiction are writers of hypertext fiction, a number
that dwindles into insignificance alongside those who recently
plunked copies of Charles Frazier's *Cold Mountain* onto their Visa
cards. While *Cold Mountain* may well have been stacked alongside a
million bedside tables during 1998, however, the people reading *Mrs*

Dalloway, let alone *Ulysses* or *Gravity's Rainbow* or any of the works by topographic or "difficult" writers, are mostly writers themselves, professors of English, or graduate students. And perhaps not even graduate students, as a student of mine once noted: Yeah, he'd read *Ulysses*—just not personally. If the distance yawning between the best-seller lists and the vestigial remains of the literary canon still dictating the Works That Count on university syllabi has not already brought home just how varied readers' tastes and habits are, surely a quick glance through inventories at Barnes and Noble or Amazon.com would forever destroy the myth of the Reader, that singular, educated entity who once queued at the docks or outside bookstores awaiting shipments of the latest from Dickens, Henry James, or Saul Bellow.

As we saw in chapter 1, readers enjoy the trancelike spell, immersiveness, and ability to screen out the buzzing world around them that are the hallmarks of ludic reading only when they are reading books that are undemanding, immersiveness existing in inverse proportion to the complexity of the characters and prose. Even if we disregard the nostalgia for the now-vanished educated reader who never existed in significant numbers, a deeper irony still underlies Birkerts's and Miller's horror at the postmodern interactive barbarians at the gate: their educated reader exists on a continuum sandwiched somewhere to the right of your average consumer of Harlequin romances at the utterly pedestrian end of the scale but far to the left of readers tackling the likes of *Ulysses* on the difficult, demanding end. Simpler, highly conventionalized texts more completely absorb any reader's cognitive capacity for comprehension than difficult ones—with the depth of readers' immersion in fiction inversely proportional to the complexity and originality of the reading matter.[3] Demands made on readers grappling with *Ulysses* require frequent pauses and regressions, breaking the "readerly enslavement" so valued by Miller and Birkerts alike. Conversely, highly conventionalized plots, stereotypic characters and settings make for an ease and more even pace of reading that absorbs readers' cognitive capacity more completely, leading to the absorption and trancelike pleasures of ludic reading. Far into the nineteenth century, reading fiction was seen as the equivalent to furtive sessions with the sherry, probably because readers became "lost" while reading light fiction—the equivalent of today's genre or mass-market fiction: "The effect of inordinate addiction to light reading . . . came under the head of 'dissipation,' and to read novels, as to drink wine, in the morning was far into the century a sign of vice."[4]

The very reasons why Austen, Dickens, and Dostoyevsky seem such thoroughly beguiling bastions for humanists everywhere is the ease with which readers can lose themselves within texts with enough of the ingredients common to conventional plots and stories to make for the entranced, immersive experience that satisfies our core desires for reading. Yet these texts also contain enough superficial and local complexity to render them interesting enough to engage readers more profoundly than genre novels and slick best-sellers while reading—and are sufficiently unchallenging to not demand the pauses and rereadings of most avant-garde and post-modern fiction. These features alone, however, are not sufficient to have earned novels by the likes of Austen and Flaubert regard from critics in every new generation, while the best-sellers of the last decades have vanished from reading lists, print, and memory.

In her study of the contribution of artificial intelligence to narratology, Marie-Laure Ryan distinguishes between vertical and horizontal motivations that drive narratives, two different relationships between the intentions that power and drive characters and plots alike. Motivation is vertical when it justifies the plot through ideas that transcend the narrative events, appealing to larger, social and philosophical categories familiar to readers but tangibly outside the confines of the story. In horizontal motivations, however, while some events might be subordinated to others, justification, ultimately, remains entirely within the plot, nestled securely and tidily inside its temporal sequence—another feature that may also account for the popularity of the highly conventionalized novels of mass-market and genre fiction that rely almost entirely on horizontal motivations, as well as of the novels that constitute the mainstream of literary fiction.[5] Horizontal motivation also makes plots more interesting because it invokes our tendency to perceive events in terms of causation, as mentioned in chapter 3, which may well account for one of the primary reasons we read for pleasure. If narrative, as Bruner has suggested, is about "the vicissitudes of intention,"[6] it is also, as historian Hayden White argues, about seeing events "display the coherence, integrity, fullness, and closure . . . that [in life] can only be imaginary."[7] The ways in which interactive narratives map and yet do not map onto this concept speak eloquently to potential for future development in hypertext fiction and digital narratives alike. And to the reasons why we listen, read, or watch fictions in any medium unfold, climax, and resolve for no purpose aside from the unalloyed pleasures they give us.

Narrative Schemata: The Changeless Story

[Narratives] seem to satisfy a universal craving for a unified, closed, and imaginary analogue to life in an open-ended and accident-prone world.

—Bert O. States, *Dreaming and Storytelling* (1993)

Not surprisingly, in the early stages of any medium, few genres exist. During each medium's incunabular phase, moreover, a small number of genres flourish, wither, and die: between 1450 and 1500, the 20 million incunabular print texts produced included ballads and chapbooks, vulgarized versions of chivalric tales—old and familiar forms easily ingested by poor readers who passed them on, hand to hand, until they disintegrated. Of these early forms of print fiction, no current descendants survive, entire genres wiped out by the advent of penny periodicals in the late eighteenth century.[8]

More recently extinct in our own century: the kinetoscope shorts that represented fodder for nickelodeons, with the radio serials that once dominated the airwaves hanging on at the top of the endangered list, represented mostly by Britain's *The Archers,* a nearly sixty-year-old relic that predated television. Strikingly, the continuity of what we might call the macroplots of radio serials—questions regarding the life choices, health, crises, and motives of the characters that extend over weeks, months, or even years—temporarily vanished from primetime television during the sixties and early to mid-seventies when *The Munsters, The Brady Bunch, All in the Family, Barney Miller,* and even dramas like *The Waltons* alike focused mostly on microplots, dilemmas easily described, pursued, and resolved within each program's thirty-minute or hourly slot. Whereas radio producers hoped to keep listeners tuning in each week to discover the outcome of a decision or even the fate of particular characters, television producers probably hoped viewers would return to programs because they identified with the characters, the particular milieus in which they lived, or even with the look of series like *The Avengers* and *Mission: Impossible!* With the development of *Hill Street Blues,* however, producers returned to macroplots as valuable devices that ensured viewers returned to watch the show, pursuing resolutions to macro- and microplots alike each week—a formula for designing, writing, and producing a series that has since become a staple of network television.

Of course, macro- and microplots have long been the bricks and mortar of stories, dating back even to Homeric epics. In *The Odyssey,* Odysseus struggles to return home safely to Penelope—the macroplot—battling against obstacles like the Cyclops and shipwreck on Ogygia—microplots that may also impact on the macroplot. While microplots involve their own smaller dilemmas, climaxes, and resolutions, their complications generally explicitly or implicitly threaten successful resolution of the macroplot. If Odysseus's crew caves in to temptation and slaughters Helios's cattle, their ship will be wrecked, and Odysseus may be left stranded—or possibly even dead—bringing the story of his struggle to return home to Penelope to the deadest of dead ends. Likewise, a microplot in *ER* might involve detoxing a drug-addicted newborn, a potentially dangerous process that brings a full-blown investigation down on Doug Ross's head. Carol Hathaway's reaction to Doug's quandary—will she back him up or censure him?—both affects the trajectory of the microplot and nudges closer to resolution one of the larger elements in the macroplot extending over the lifetime of the series, involving Doug's philandering and inability to make a steadfast, long-term commitment to her. As much as they might be bewildered at the outward trappings of the stories in *ER*, the audiences who once listened to Homeric rhapsodes would recognize the plot schema represented by the series—because it represents a story schema that is as ancient as stories themselves: whether spoken, written, recorded, filmed, created with Photoshop and RenderMan, or posted on the World Wide Web.

At one time or another, everyone from linguists like Vladimir Propp and A. J. Greimas to psychologists of reading such as Walter Kintsch and Teun van Dijk have attempted to explain how narratives work. Few theorists, however, have been able to describe *why* narratives work, and in particular why, say, the Oedipus plot can resurface in several hundred guises (including its starring role in Freud's case narratives) over thousands of years without its core appeal ever being exhausted. To begin to answer these questions, we can examine the ways in which one of the most sophisticated examples of digital narratives represents, surprisingly, a veritable paradigm of classic storytelling, relying on rules authors use for telling stories and processes readers use for comprehension as fleshed out by Robert de Beaugrande and Benjamin Colby—rules that enable us to isolate the features of stories that readers consider well told or interesting.

At its core, every story is about characters' plans to attain goals—even when the particular goal may be returning to the state the character enjoyed at the very outset of the story, prior to tackling the steady stream of opportunities, complications, or calamities that throw the plot into gear. Often, plans and goals conflict with one another, even when a single character holds them. For example, in the digital narrative *The Last Express,* protagonist Robert Cath wants to avoid attracting attention while aboard the Orient Express because he seems to be hiding out from the police. Yet, when his friend and compartment-mate Tyler Whitney is murdered just after the train pulls out of the Gare de l'Est, Cath must balance his desire to remain on the train invisibly—given reasons he has for avoiding the police himself— with his wish to discover the identity of his friend's slayer.

Not surprisingly, the plans and desires of one character frequently contradict or clash head-on with the intrigues and ambitions of others. Cath's desire to discover both Whitney's killer and what his pal might have been up to just before his murder conflicts in a highly complex and roundabout way with Serb passenger Milos Jovanovic's goal of securing a shipment of guns and ammunition from German industrialist August Schmidt, munitions the Serbs need to free Serbia and Croatia from the grip of the Austro-Hungarian Empire. August Schmidt's price for the guns is Prince Kronos's gold-filled suitcase, which Kronos will trade only for the jeweled firebird egg stolen from Tyler Whitney by one of the twenty-nine passengers on board the Orient Express. Of course, each local—or micro—goal potentially conflicts with other microgoals. Cath must play along with both Schmidt and Kronos, pretending he has something to trade with each of them until he can recover the firebird or steal the gold or both. If he fails to deliver the guns to the Serbs, they will probably kill him; if he delivers the guns to the Serbs, they will probably commit terrorist acts against the Hapsburgs still controlling the empire—something that some readers know will result in the assassination of Archduke Franz Ferdinand, the tinder that ignited the long fuse leading to the outbreak of World War I. The Serbs' plans, a famous and attractive violinist's secret spying mission, even seven-year-old François Boutarel's fascination with whistles and beetles, all complicate, threaten, and—because *The Last Express* is an interactive narrative—potentially or actually terminate the macroplot and the reader's experience of the story—if only temporarily. Plans, conflicts, goals, clashes, and rewards are the stuff from which everything from *The Odyssey* to *The Last Express* and "Twelve Blue" are made.

Beaugrande and Colby's relating storytelling rules to processes of comprehension is unusual because of their definition of interesting and enduring stories. Goals and actions, states and events cannot be so obvious that their outcomes are certain or simply retrace the normal outcomes familiar to us from life. For us to be drawn into narratives, the relationships among characters, actions, results, and reactions must be uncertain.[9] All participants in narratives—the narrator, the narratee or audience, and the characters involved in the plot's intrigues and actions—spend much of their time predicting, obliging the narrator to "outplan the audience at least part of the time to keep the story from becoming predictable and boring" (49). Mystery stories derive much of their tortuous twists and rogues' galleries of suspects from this need, leading the narrative to encourage readers' misdirected suspicions in every direction possible until the stories' climax. Even in other, less intricately plotted genres, however, readers learn in detail about characters' aims and plans, leading them to anticipate the probable outcome of the conflicts that lie just ahead by relying on their own experiences. The more intricate and difficult the problems, the "greater the energy and the deeper the processing expended on story comprehension" (49). When in "Twelve Blue" Javier and his daughter, Beth, visit the hotel where its owner, Ed Stanko, possesses the only existing photograph of Javier's grandmother, we know enough about Stanko to realize that his character is sufficiently bitter, twisted, and stunted to make it unlikely that he will so much as let them see her photograph, let alone surrender the portrait to them. Joyce's narrative, however, turns our predictions back on us, blunted. Instead of a violent confrontation between Stanko and Javier, we discover Javier and his daughter arriving at the hotel in time to encounter Stanko's mad tenant, Eleanor, freshly daubed with her now-dead landlord's blood.

The twists in "Twelve Blue" are unexpected and thus heighten our pleasure in the narrative as we witness the author outplotting us, urging us to guess, then revealing how our guesses fall short of reality. But what of the genres that rely on a slender array of story types, or, even, stories—like the Oedipus plot—that have been recycled for millennia? How can we, who know its intrigues and revelations so well, still take pleasure in its unfolding, if so much of our pleasure is bound up in prediction, anticipation, and discovery?

Beaugrande and Colby venture two possible explanations why recycling stories need not remove uncertainty from narratives. First, global and local processing of information—for example, recognizing and remembering the types of goals and actions common to characters in thrillers involving espionage—draws attention away from the particular details of the likes of Robert Cath's goals and actions aboard the Orient Express. They occur on a different cognitive level from our processing information about Cath's status as a twenty-nine-year-old American and amateur agent provocateur who may or may not be on the lam from a few botched intrigues of his own before he boards the train in Paris:

The knowledge of global structures of a narrative might not be on the same level of processing depth. . . . Interest is upheld during repetitions of the same narrative because the audience predicts only global data, and rediscovers local data each time. . . . [E]nduring narratives—and perhaps art objects of all kinds—manifest inherent structural complexities whose processing demands, even after repeated exposure, remain above a certain threshold of cognitive storage abilities, and yet below the threshold where ongoing processing would simply break down.[10]

Another complication—or reason we never tire of some stories—lies in the significant energy readers expend in anticipating the consequences of actions, events, and reactions throughout a narrative.[11] Further, as so many stories invite readers to anticipate murder, mayhem, love, and death, readers may persist in anticipating disastrous alternatives at the end of each narrative junction or strand—partly, as some critics have claimed, to satisfy an innately human need for intense excitement, leading them to indulge in romance, violence, and death vicariously.[12] Identifying with a character—however fleetingly—and anticipating a Jason or a Freddy lurking just around the corner can "awaken the same sort of anxiety people undergo when recalling their narrow escapes in real life. In retrospect, people are safe just as narrated protagonists are known to be safe after earlier narrations; but tension still arises from mental reconstruction of what might (or even ought to) have happened."[13] Another reason why outplotting the reader's expectations is instrumental to telling a satisfying story: we turn to narratives to slake our thirst for danger, excitement, adventure, and to reassure ourselves that the world is, after all, an orderly, secure, relatively peaceful, and, above all, mostly predictable world.[14] Narratives resolve these two apparently irreconcilable longings by placing the violence, destruction, and danger within highly conventionalized forms that recover for us intentions, emotions, and many inward states normally inaccessible to us, at the same time they also provide the entire package in stories that let us observe the neat causal sequences, the well-defined beginnings and endings forever denied us in life. Ed Stanko's murder in "Twelve Blue" is the sort of item that occasions a flurry of news stories that dance around the central conundrum of any homicide—the why—without our ever gaining insight into the intentions that flickered through the killer's mind when she picked up the knife. In Joyce's Web-based fiction, however, we can enter Eleanor's muddled, frenetic thoughts, discovering in their tangles her conviction that Stanko not only fathered her baby but may also be responsible for its death.

Millennium Story: Hypertext Fiction and the New Realism

> [T]oday's most successful interactive artists ultimately see
> interactivity as an evolutionary (rather than revolutionary)
> step for storytelling.
> —Brent Hurtig, "The Plot Thickens" (1998)

Strikingly, in "Twelve Blue," as in films like *Nashville* and *Short Cuts*, there is no macroplot, only a myriad of microplots that touch each other physically, coincidentally, metaphorically, but never connect causally in a single overarching plot that brings the story into existence and offers the resolution that signifies its completion.[15] Readers, instead, confront resolutions to some of "Twelve Blue"'s microplots, guided visually through a graphic interface that stands in for the narrative's macroplot: a drawing of twelve brightly colored threads representing "Twelve Blue"'s narrative strands, stacked horizontally in a frame divided into eight bars that symbolize the narrative's temporal axis (see fig. 10). As readers move through the text, they see only the segment of the graphic pertaining to their temporal place within the hypertext. When the threads arc toward, touch, or veer away from each other, the stories represented by each strand follow suit, although the narrative strand containing the story of a drowned deaf boy and the fate of his corpse drifts across the other narrative threads, seeping into other plots—most notably, the hallucinations of the dying Ed Stanko—until it surfaces as the colored thread curves upward to the very edge of the frame (see fig. 11).

While many postmodern writers have traded macroplots for a different set of complications and effects—those of the difficulties and dangers of narration, of the telling of stories itself—Joyce introduces an entirely different element into the writer's arsenal of plot, character, narrative, cause, and effect. The image from which "Twelve Blue" partly derives its name corresponds to the revolving center of the text—not a segment of text or a climactic instant, but a graphic, the image of bright threads swimming against a field of blue. Appearing alongside each screen of text, the threads trailing across every segment of the graphic act as tangible guides to the trajectory of micro- and macroplots alike, symbols of the ingredients of each narrative strand, as well as the primary mechanism by which readers move from link to

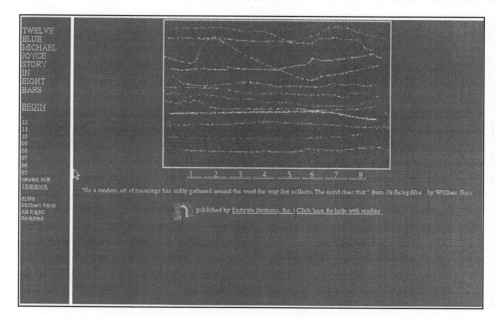

Fig. 10. Narrative skeins made visible: the graphic central to the "Twelve Blue" navigational interface.

link as they inch along the horizontal axis of the image, clicking on one of the colored threads.[16] Chief among the distinct technical differences between the World Wide Web and earlier media of representation is its ability to link image and text seamlessly, enabling a uniquely close interplay, even a marriage, between image and narrative, between symbol, plot, and the surface of the story, one that offers glimpses of striking possibilities for the future of hypertext fiction.[17]

Further, Joyce's carefully scripted links bring us the voyeur's point of view, supernaturally privileged from time to time as it drifts from consciousness to consciousness, dipping briefly midstream into the thoughts of a mad woman, the experiences of a drowning boy and the fate of his drifting, decaying corpse, the early flirtations between a couple, the final hallucinations of a dying man. We have moved backward again to the overheard snatch of conversation, the nugget of story buried amid the detritus of everyday lives, all the tiny threads of other lives that briefly brush against ours as we race through our days, immersed in our own micro and macrostories. Perhaps this truly is the "New Realism,"[18] a fiction that, as Joyce has imagined,

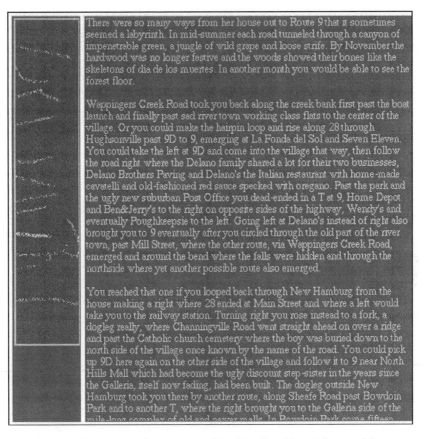

There were so many ways from her house out to Route 9 that it sometimes seemed a labyrinth. In mid-summer each road tunneled through a canyon of impenetrable green, a jungle of wild grape and loose strife. By November the hardwood was no longer festive and the woods showed their bones like the skeletons of dia de los muertes. In another month you would be able to see the forest floor.

Wappingers Creek Road took you back along the creek bank first past the boat launch and finally past sad river town working class flats to the center of the village. Or you could make the hairpin loop and rise along 28 through Hughsonville past 9D to 9, emerging at La Fonda del Sol and Seven Eleven. You could take the left at 9D and come into the village that way, then follow the road right where the Delano family shared a lot for their two businesses, Delano Brothers Paving and Delano's the Italian restaurant with home-made cavatelli and old-fashioned red sauce specked with oregano. Past the park and the ugly new suburban Post Office you dead-ended in a T at 9, Home Depot and Ben&Jerry's to the right on opposite sides of the highway, Wendy's and eventually Poughkeepsie to the left. Going left at Delano's instead of right also brought you to 9 eventually after you circled through the old part of the river town, past Mill Street, where the other route, via Wappingers Creek Road, emerged and around the bend where the falls were hidden and through the northside where yet another possible route also emerged.

You reached that one if you looped back through New Hamburg from the house making a right where 28 ended at Main Street and where a left would take you to the railway station. Turning right you rose instead to a fork, a dogleg really, where Channingville Road went straight ahead on over a ridge and past the Catholic church cemetery where the boy was buried down to the north side of the village once known by the name of the road. You could pick up 9D here again on the other side of the village and follow it to 9 near North Hills Mall which had become the ugly discount step-sister in the years since the Galleria, itself now fading, had been built. The dogleg outside New Hamburg took you there by another route, along Sheafe Road past Bowdoin Park and to another T, where the right brought you to the Galleria side of the mile-long complex of old and newer malls. In Bowdoin Park some fifteen

Fig. 11. No matter where you read in "Twelve Blue," the graphic to the left of the text orients you within one of the narrative's eight time frames, displaying the narrative skein you've selected relative to other strands.

promises to close the gap between the fragmentary experiments of language and narrative which have characterized so-called literary or experimental fiction and the distinctly segmented consciousness of a larger audience who, from moment to moment, settle upon meaning for their lives in the intervals between successive accounts of their own or others' lives in several media. . . . [It is] a narrative which can make sense of life as it is lived outside the regime of nextness . . . hypertextuality somehow represent[ing] the ordinarymindedness . . . of most people's lives.[19]

A Little World Made Cunningly:
Digital Narratives and the New Realism

> To give the player the feeling of being in a populated world,
> we had to make sure that we knew what every character was
> doing at all times, just in case the hero of the story wandered
> into one of them. Even though the conductor had a very
> small part, we had to script out a two-minute conversation
> about politics he was having. And when you do that for 40
> characters it becomes a huge amount of writing—far more
> than a traditional script, and it's all in the background, but it
> gives [the story] a richer texture.
> —Jordan Mechner, creator of *The Last Express* (1998)

> Who the hell wants to hear actors talk?
> —H. M. Warner, Warner Brothers (1927)

In the essay in which he introduced and named the New Journalism, Tom Wolfe attributed the power of the modern novel to the four devices he felt gave it immediacy as well as the capacity to both move readers and absorb them: a narrative rendered scene by scene, eliminating the need for an omniscient narrator or bridging narrative; dialogue recorded in full because "realistic dialogue involves the reader more completely than any other single device"; the third-person point of view used to present each scene to readers; and the minute recording of the gestures, furnishing, dress, behavior, and idiosyncrasies that enable readers to understand something of the protagonist's interior life, desires, goals.[20] One of the most admirable accomplishments of the New Journalism has been its relatively recent influence on nonfiction, for example, decidedly nonjournalistic works like Julia Blackburn's *Daisy Bates in the Desert*, Lawrence Weschler's *Mr. Wilson's Cabinet of Wonder*, and John Demos's *The Unredeemed Captive*, works that harnessed the power of narrative fiction to bring readers closer to moments that have receded into the distant past, to bring us closer to realities long gone that remain to us mostly unimaginable—territory that the best digital narratives have now also colonized.

New Journalism's second legacy has gone mostly undetected: it is quite possibly the most successful low-tech bid for realism since the Greeks discovered perspective in art. One of the great paradoxes of realism is that you need as much technology as you can muster to summon it up because the old realism that seemed perfectly adequate when representation was limited to words or paintings suddenly seems obsolete when artists begin playing with daguerreotypes, or kinetoscopes, or steadicams, or Photoshop. The next wave of reality after the ebb of the one presently breaking may involve suits wired for simulated touch, temperature, impact, and head-mounted devices for replacing the world before our eyes with one mostly fictitious, a jazzed-up version of the feelies Aldous Huxley described in *Brave New World*, sans the knobs and with a lot more circuits. In the peculiar, paradoxical way of things, as both Walter Ong and Marshall McLuhan long ago pointed out, technology has a way of making us more fully human. And of making the aesthetic object that promises to deliver us a facsimile of the world seem a little more real with each layer of complexity that you insert between the object and the world it purports to capture and deliver to us.

The problem with cinema that even Scope and Dolby and IMAX can never resolve, however, no matter how technologically sophisticated we make the medium, is that cinematic protagonists are almost invariably doing cretinous things: venturing into dark cellars when the electricity has been cut off, running in six-inch heels and tripping before flesh-eating dinosaurs, poking around in the sock drawers of neighbors they suspect of bumping off their spouses when their neighbor ducks out for the morning paper. Even while we reluctantly enjoy the suspense of wondering if the hero or the victim is going to make it to the next reel, we mostly believe that, whatever happens, they've got it coming to them. *We* would never have walked into the house, let alone traipsed down the basement stairs or stumbled and fallen down helplessly, wailing, before we are mown down. Digital narratives have long promised that we could come up with our own strategies, our own solutions, even some neat footwork while fleeing the horror of the moment. But games like *Obsidian, Myst, Midnight Stranger,* and *Douglas Adams' Starship Titanic* mostly have failed to deliver on them, confining users to sword and gun play and battles, solving puzzles, sniffing out obscure clues, killing trolls, and manipulating the myriad clichés that have marked the medium out mostly as turf for preteen boys and adults with the sensibilities of preteen boys. Try this showdown on for size, sonny. Let's see how fast you are with that joystick.

Until *Myst,* digital narratives remained the equivalent of the pulpy end of genre fiction—a very high tech version, extremely costly to produce, of the romance novel, a genre that recycles its characters, heroes, heroines, and plots with an assiduousness that would have garnered praise from the Sierra Club if the resources in question were anything other than purely imaginary. As savvy editors and publishers in the romance game doubtless know, the stereotypes are window dressing, something to drape over a plot that provides its romance-starved readers with the paper approximation of a quick fix. Likewise, when digital narratives featured scenes or characters, these served as mere conduits to hustle the player along to the next test, a coming-of-age ritual reenacted by millions of joystick-wielding teens, who mostly wanted to leap dungeons and precipices, fence, fire pistols, and wage battles against the clock, the threat of death, or the high scores accumulated by the runty kid down the block. Adult versions of the teenage digital narrative include *Midnight Stranger,* where the goal of finding an extraterrestrial object is the pretense for coming on to a succession of women, practicing the old in-out, in-out some dozen times with a constellation of attractive females after they invite you home with them, and battering a hapless drunk for the sheer fun of it. Some day, sonny, this will all be yours.

Myst bestowed respectability on digital narratives because its environment was so globally rendered, complemented by what appeared to be languid lap dissolves, eerie, isolated sounds, and a complex array of clues and puzzles that took users weeks, sometimes months, to unravel. Yet, if *Myst* was immersive, it was also a far cry from even the narrative richness of your average paperback plucked from the fantasy section at the local bookstore. It lacked characters more substantial than the mere flickering faces of Atrus and his sons, fragments of their conversation, and the odd, isolated scrap of writing knocking around on the lawn or in the library. Set conveniently on an island, *Myst*'s environment was highly, if artfully, limited, as was the corresponding dearth of things its readers could do on the island. And the reader as the protagonist had no looking-glass self in the text, no defined role within the narrative that blurred or rushed into focus depending on who you were interacting with, because there was no one around. You either successfully completed the tasks, the puzzles to solve and artifacts to recover, or you stayed stalled ignominiously in the Selenitic Age—or, worse still, traipsing around Myst Island thrashing on shrubs and colonnades uselessly with your cursor and mouse, praying for clues, the answers to puzzles, or the imminent publication of the *Myst* strategy guide.

Even when characters address you in narratives like *Gadget* or *Starship Titanic,* however, your role is entirely extranarrative. You are entering a world expressly to pick up pieces, sort things out, and generally restore order, and, to accomplish these tasks, you have to abide by certain highly specific and particular rules. Until recently, the game has never lurked more than a few millimeters beneath the digital narrative's veneer because, without it, the narrative lost most of its raison d'être. Either that, or, without the game challenges and overall framework, producers needed to provide more richly detailed environments, more branching possibilities that spawned more plot possibilities to be realized and rendered, and characters with appreciable depth and complexity, all of which can seem needlessly expensive when your audience has been made up of fourteen-year-old boys mostly hankering to use their joysticks.

The digital narrative kill-or-be-killed scenario has, however, received recent nudges and prods, finally making its first strides outside video arcade mode and into realms more familiar to cinemagoers and readers of fiction. First, Shannon Gilligan's Multimedia Murder Mystery series seized on the mystery and breathed fresh life into it by placing readers squarely in the investigative driving seat. A few decades of TV cop shows have given most readers an easy familiarity with police procedures and a hankering for a bloodless and stressless flirtation with directing an investigation themselves: scanning the coroner's report, analyzing blood spatter patterns and fingerprints, the usual detritus of death scenes.

Using the conceit that six hours, on average, elapses between the discovery of a crime and the arrest of the alleged perpetrator, digital narratives like *The Magic Death* and *Who Killed Brett Penance?* place all the usual tools of homicide investigations in readers' hands, including a sidekick who guides you through the ropes, hints at recommended strategies for interviewing suspects, and occasionally gives you the skinny on less-than-forthright suspects. Your interaction, however, is limited to the tools of the detective trade—investigating the crime scene, reading the coroner's report, and interviewing suspects, limited to questions in your notebook that you can choose to ask each one. And your choices always matter, because each interview question, each test, each request, eats away at your allotted six-hour maximum with chunks of time commensurate to the amount and importance of the data you receive. Since interviews with suspects merely trigger well-acted video clips, moreover, the characters just respond to the question, or to the generic role you are temporarily filling, the detective on the trail of an arrest warrant, and not to any modicum of personality the detective may possess—a far cry from the long line of colorful detectives that extends from Philip Marlowe to Patricia Cornwell's Virginia West. Still, Gilligan's series considerably extends the narrative complexity and pleasures of a single mystery narrative by offering in *The Magic Death* three different perpetrators to nail and in *Who Killed Taylor French?* and *Who Killed Brett Penance?* three entirely different crime scenes, three perpetrators' modus operandi to analyze, and, of course, three culprits. In the age of obsolescence, remarkably, Gilligan's narratives considerably extend the shelf life of your typical mystery, which readers usually consume quickly and which is seldom suitable for rereading unless you have a terribly short memory, because the resolution of the print mystery's macroplot destroys most of the limited pleasures mysteries—with their sketchy characters and mostly pedestrian prose—offer.

With the debut of *Titanic: Adventure out of Time* readers at long last enjoy a richly detailed environment to explore and a compelling cast of characters, a world limited by the ideal conceit: a ship in the middle of the ocean with a scant five or six hours before she plunges to the bottom. In *Titanic* you begin the narrative with a modicum of identity—a few postcards, a tacky flat, and a career that, judging from the correspondence littering the drawer, went into permanent eclipse with the sinking of the *Titanic*. Once you enter the shipboard narrative itself, you assume the mantle of British secret agent, society gent, and something of a genteel rake, if the protestations of Lady Georgia are to be believed. Perhaps more important, however, your actions and reactions to the other bodies populating the ship ease or interfere with your general mission of recovering the stolen rare original of *The Rubáiyat* and a notebook recording the identities and whereabouts of the Bolsheviks. Brush off tireless and tiresome society gossip Daisy Cashmore when she asks you to discover a fellow passenger's identity, deal with Willie von Haderlitz in strictly hostile terms, and you might be in for a very long and fruitless night indeed. For once in the medium, character truly *is* action, and action, character.

Still, for every scrap of freedom the reader enjoys in a digital narrative, programmers and designer sweat hours and thousands of lines of code, and producers, more importantly, sweat the number of digits in their outgoings columns. Which is one reason why *Titanic* restricts its readers' opportunities for interacting with characters by providing them with a multiple-choice list for salutations, responses, and challenges alike. Frequently, the reader's choices are unobtrusive: assenting to an opinion, accepting an offer to play cards, declining another drink with Georgia's boozy, snooty husband. More often, unfortunately, they restrict characters to the tics and drives necessary to fulfilling their appointed roles within the micro- or macroplot scenarios in which you encounter them. Ask Officer Morrow point-blank if you can stick your head around the door to the telegraph shack—a newfangled creation in 1912—and he will send you on your way. Likewise, if you choose the remarks that offer him a drink or volunteer that you believe war unthinkable, he also sends you packing. Like virtually all the characters crowding *Titanic,* what there is of Morrow's character is strictly a function of his responses to your multiple-choice rejoinders, an improvement over the people-less universe of *Myst,* but a far cry from characters E. M. Forster might have described as "round."²¹ And, as I noted in chapter 1, if you decline to pick up clues that include Russian dolls and telegrams and negatives lying on darkroom trays, you are forever condemned to trundling around a ship drained of characters, presided over by a clock that remains stubbornly stopped until you concede defeat and go back to assembling a treasure hunt's worth of clues, like a good player should.

Still, the pleasures of immersion in this lavish, belle époque environment, fabulously opulent, famously unrecoverable, are hefty, heightened by faithful renderings of *Titanic's* interiors and haunting musical accompaniments that complement each area of the ship with different musical themes. The recovery of the Gilded Age with its lavish, privileged swank and settings is, after all, a valuable commodity to late-twentieth-century audiences accustomed equally to a sense of time's profound scarcity and to the bland homogeneity of airports, shopping malls, and hotels the world over—it is partially what catapulted the film *Titanic* to success, what still makes a transatlantic crossing aboard *Queen Elizabeth* 2 the sort of event travelers describe reverently. It is also, potentially, one of the singular pleasures digital narratives can deliver—an invitation to experience a simulacrum of a world that vanished forever with the onslaught of the Great War. And, not coincidentally, this same *Time Machine*–like feature figures heavily in the appeal of Jordan Mechner's *The Last Express,* a digital narrative set in 1914 aboard the Orient Express during a three-day journey from Paris's Gare de l'Est to Constantinople.

Readers entering *The Last Express* encounter animated sequences as the Orient Express idles at the platform and a nervous Tyler Whitney scans the station, eyeing the clusters of gendarmes watching the trains. Then, as the train chugs away from the station, a motorcycle races alongside it and Robert Cath neatly and unobtrusively boards the Orient Express with a leap from the behind the motorcycle's driver onto the train. Once he's aboard, however, the opening animation ends and Robert Cath is more or less yours—or, more accurately, you are more or less him: able to direct his hands, feet, and voice with a freedom that is quite naturally limited by Cath's own tendencies toward quick-wittedness, adventurousness, and occasional sarcasm. Move your cursor and Cath follows. When your cursor passes over a door or object, it turns into a surrogate for Cath's hands. When opportunities for conversations with other passengers and crew arise, your cursor morphs into a cartoonlike conversation balloon.

Your arsenal of actions and movements feels natural, unusually lifelike because Robert Cath is no tabula rasa, a blank space inviting readers to insinuate themselves into the narrative. As you learn from encounters with fellow passengers and some digging in the bag Cath carries with him, he's a man with a past that remains beguilingly murky, suggested through telegrams and newspaper clippings and overheard conversation. This is also no bland protagonist who dutifully collects objects and whose conversation you never hear. Decide to head toward Prince Kronos's private railway car early in the narrative, and you will listen to Cath trade barbs with Kahina, the prince's bodyguard. Direct Cath to approach Anna Wolff on her own in the dining car, and you will watch how he handles a brush-off. Since your participation in the narrative is also directive, purpose-driven, Cath's actions unfold in purposive sequences: click on Whitney's duffel bag and it opens, but the narrative leaves what you riffle through and pick up mostly to your choice. And the grain of your interactions is fairly fine, allowing you to explore the train, speak with other passengers, pick up newspapers or a conductor's sketchbook, or sleep according to your particular purpose or whim. Of course, you must react adequately to challenges as they arise, quite naturally, in a narrative that begins with a murder and features a cast of twenty-nine characters with conflicting goals, including two clutches of terrorists, an assortment of spies, a hyperactive and obnoxious seven-year-old, and a good half-dozen potential murderers. Fail to hide Whitney's body adequately or to hide from the police yourself, and the narrative halts as if somebody had hauled on the emergency cord aboard the train—an action you can also take, particularly if you have been nosing your way through *The Last Express* and are searching for a quick out to end the evening's entertainment, since hauling on the emergency cord, not surprisingly, also ends the narrative rather efficiently. The screen quickly whites out, then irises in on an extract from the diary kept by Rebecca West surrogate Rebecca Norton, a diary in progress on which you can keep tabs through periodic snoopings in Compartment E, the current entry summarizing the untimely departure of Robert Cath from the Orient Express, seen from the perspectives of Norton and her companion, Sophie de Bretheuil.

As interfaces inviting interaction go, *Last Express* represents an evolutionary leap beyond even the ostensibly open-ended input for exchanges with bots in *Starship Titanic* and its PET interface that

Starship Titanic so strenuously attempts to incorporate within the confines of the story itself, but which remained mostly a tool for navigating through the ship and manipulating objects. Each direction provided by readers in *The Last Express* triggers entire sequences, so that the interactivity meshes neatly with the core story schematic of situation-event-action-reaction identified by Kintsch, Beaugrande and Colby, and other narrative theorists. Click on the body of Tyler Whitney sprawled on the floor of compartment 1, and you will watch Cath strain to pick him up and lay him on the banquette. Leave the body on the banquette, and the conductor will discover him, haul on the emergency cord, and summon *les gendarmes.* Hide the body in the made-up bed, and you will need to scuttle back quickly to the room before the conductor arrives to make the bed up for the night, so that you can toss Tyler's body out onto the tracks. Or, if you point to the window, open it, and heave the body in that general direction, your pal's corpse flies out onto the tracks between Paris and Epernay, the tidiest solution—and one that leaves you with a bloodstained jacket. Point to the jacket hanging above the banquette, however, and Cath seizes it, swaps it for the now bloodstained jacket he wears, and tosses his own bloodied jacket out the window and onto the tracks. When moving, your point of view is fixed squarely within Cath's perspective; when directing his hands or prodding him into conversation, you assume the third-person limited perspective on Cath, giving you at once the voyeuristic pleasures of dipping into another's consciousness, the fun of seeing Cath doing what you have commanded him to do, and of seeing yourself as others see you. The perspective reminds us of cinema's interplay between first-person and third-person perspectives, without, however, the strictly voyeuristic role we fill as we watch films. In *The Last Express* virtually every decision you make not only fleshes out the lost, privileged world represented by the cross-section of society aboard the elegant Orient Express but also determines your course along a narrative that branches repeatedly.[22]

Not surprisingly, many of the branches are tracks to failure. Open the door to the conductor before you have had a chance to dispose of Tyler, and he will see Whitney's body, stop the train, and have you arrested. Fail to hide when two gendarmes search the train at Epernay, and you will be arrested for the murder of Whitney, whose body has been recovered from the tracks where you have chucked it. Botch delivery of the suitcase to August Schmidt, and he removes his cache of weapons from the train in Vienna, leading the Serbs to kill you in a rage as they watch their planned revolution carted away by porters. Whereas digital narratives formerly punished wrong moves with speedy deaths or successions of doors that refuse to open and characters who cannot be approached, *The Last Express* enables you to back up and pursue different tracks that invite you to continue eavesdropping, rummaging through briefcases and under pillows in empty compartments, even liberating Anna Wolff's dog, Max, from the baggage car, so he can menace Kahina as she breaks into Wolff's compartment while Wolff plays the violin for Prince Kronos in his private car at the rear of the train. Just when you believe you made all the logical choices, however, you can discover in Vienna that swapping the jeweled firebird egg for Prince Kronos's suitcase bulging with gold can strand you in Vienna with an unusual ending for a would-be sleuth or hero. The last scene in this particular version of the narrative features Cath sipping cappuccino in a sidewalk café while straddling the fortune stowed in the suitcase between his knees—as Kronos hurries away with the firebird and the train puffs onward to Constantinople—leaving unsolved and unresolved virtually all of the conundrums you have encountered during the narrative.

The gamelike aspects of *The Last Express* provide your actions with purpose, with tangible repercussions for the choices you make, the options you exercise. You can, however, focus on the detailed conversations, the tics and idiosyncrasies of other passengers, the fragments of Joyce's "ordinarymindedness" that flesh out the narrative and comprise more than half of its content.[23] Entire conversations exist merely for you to eavesdrop on them without their ever relating to the macroplot's intrigues. While Rebecca's and Sophie's chatter over tea in the salon potentially exists as an opportunity for you to sneak into their room and rake through their effects, you can just as easily stay behind and eavesdrop while you scan the lead stories in the daily paper—especially since Rebecca's diary is mostly a colorful take on the other passengers aboard and your snooping around the pair reveals only what sounds distinctly like an ongoing lovers' quarrel. Norton's diary functions like a one-woman Greek chorus, only her observations are strictly limited to what Norton sees, hears, and values, and this particular chorine is immersed directly in a microplot of her own as she struggles for the affections of the capricious and mostly vapid Sophie de Bretheuil. Likewise, the book you snatch from beneath conductor Coudert's chair reveals not a passenger list that might help you in your search for the missing firebird but sketches and caricatures of passengers and crew. Linger alongside the Boutarels' table at dinner and you will hear Madame Boutarel's scathing replies to her husband's ambitions for their obnoxious seven-year-old son, François, another exchange that does absolutely nothing to further either microplots or macroplots. As you draw within earshot of each party, their overheard conversations are subtitled, variously, in Russian, French, and German—languages Cath both understands and speaks. Eavesdrop on Kronos or Mahmud and his harem, however, and you will find yourself in Cath's shoes, so to speak, listening to a stream of Arabic, sans subtitles, since Cath neither speaks nor understands that language.

Ordinarymindedness and Realism Squared

It is satisfying to switch position . . . to act in a patterned event and then later view the general pattern, like a synchronized dancer. . . . But a computer simulation offers a new extension of this pleasure. On the computer we can reenter the story and experience more than one run of the same simulation. We can . . . exhaust all the possible outcomes. We can construct a composite view of the narrative world that does not resolve into any single story but instead composes itself into a coherent system of interrelated actions. Because we increasingly see the world and even our own identities as such complex, centerless, open-ended systems, we need a story environment that allows us to make sense of them by enticing us into exploring a dense narrative world. . . . Whereas novels allow us to explore character and drama allows us to explore action, simulation narrative can allow us to explore process. Because the computer is a procedural medium, it does not just describe or observe behavioral patterns, the way a printed text or moving photography does; it embodies them and executes them.[24]

While Janet Murray bemoans digital narratives' shallow branching structures and disproportionate emphasis on visuals over storytelling (212), her concern seems overly purposive, powered solely by the satisfactions of watching actions and reactions simulated onscreen. When digital narratives dedicate resources, scripts, characters, and narrative branches entirely toward depicting actions and consequences, however, the resulting narrative may be lacking in the local and not strictly purposive detail that enriches environments and can lead readers to believe much of the narrative remains to be discovered on a second exploration. If, as in *Titanic*, the responses of characters and the rooms I can explore contain details relevant only to the potential resolution of the macroplot and its corresponding microplots, chances are my second, third, and fourth narrative run-throughs are going to feel considerably more impoverished than the first one—even making me feel as if I were merely playing the treasure hunt to satisfy my longing for the control generally denied to me in my everyday life, yet another reason for the game, battle, and joystick appeal to teenage sensibilities.

Yet if the print fiction and films we return to are generally economical at the level of plot—including the red herrings and false leads that make for enjoyable mysteries and thrillers—good narratives themselves are inherently wasteful, filled with details at every turn, the flotsam of everyday life, the exchanges heard over lunch that sketch out relationships like one binding Tatiana to her grandfather, the confessions we mostly forget made to us aboard trains, snatches of conversation like the anti-Semitic aside August Schmidt issues almost immediately after his fawning over Anna Wolff that reveal still more to us of a character we have already decided is distasteful from our brief encounters with him. Rich narrative is all about detail that accretes, containing Forster's "round" characters who, like Schmidt, might imbibe a few too many glasses of brandy and end up waltzing with Cath in the salon when his intended assignation with Anna Wolff does not play out quite as he planned.[25]

If film derives its immersiveness from its ability to depict the minute detail of life as we know it as the backdrop to story, digital narratives can square this realism by capturing minutiae that are mostly irrelevant to macro- and microplots alike. In *The Last Express* the conductors moan over the parsimonious tippers on board. A cook twists the kitchen boy's ear during prep time in the kitchens. François yanks the legs off the beetle Cath gives him when the bug refuses to follow orders during a game of soldiers. Overly loquacious George Abbott settles down uninvited alongside a brooding Alexi Dolnikov in the salon and prattles on for minutes with the occasional nanosecond pause for a rejoinder or intake of breath, not remotely perturbed by Alexi's stony silence. As Beaugrande and Colby point out, the richness of local detail in narratives cannot be fully processed and retained, leading us to reread narratives rich in small, "throwaway" details like these for pleasure, the very details that express character the way we might observe it in life, without our being aware of an author's sketching them out for us in words—yet another example of Wolfe's realism squared.[26] One of the measures of a digital narrative should be its waste, in terms of the amount of detail, characters, potential interactions, and even entire story branches that, to paraphrase Auden's declaration about poetry, make absolutely nothing happen. While realism

may entail respecting what readers know about perspective and movement from lifetimes of exposure to art, photography, and film, realism
is not necessarily about 35mm or video clips, since the look of realism
and the feel of realism can be two distinct entities, and glossy footage
cannot offset narratives where all branches lead more or less straight
to The End and a singular conclusion. And, since realism usually deep-
ens our immersion in narratives, this latest push toward a New Realism in digital narratives promises to turn the medium into a source of
pleasure, of simultaneous exploration and escape, capable of delivering the aesthetic goods as respectably as—and, for some, potentially
more pleasurably than—novels or films.

> The central engines of our mind are bent always and forever
> on the job of making stories, in large themes and a thousand
> subthemes simultaneously.
> —Philip J. Hilts, *Memory's Ghost* (1995)

Hypermedia fiction and digital narratives on disk, CD-ROM,
downloaded from the World Wide Web, or, as our battles with band-
width restrictions ease, even run in real-time off the Web—the technical specifications and look of these will morph and evolve during the
years ahead. What will not change are the things that have always
engaged us: the strings of cause and effect; generalizations about character and motivation we accrue from our study of outward dress,
manner, tics; the dense weave of micro- and macroplots; and, always,
underlying all of it, words, words, words. Contrary to the convictions
of Sven Birkerts and other Luddite critics, technology and interactivity nudge us no closer to the extinction of *le mot juste* than we were
before the invention of telegraph, telephone, television, or computers.
Beneath every interactive narrative, adventure, mystery, thriller, or
romance lie words, the scripts that can render characters lifelike and
memorable, the scenes and details that we recall long after we have
forgotten the way the thing ends. While it is possible to make a terri-
ble film from an excellent script, it is virtually impossible to turn a
hackneyed script into a watchable film, just as all the whizzy animation and three-dimensional modeling in the world cannot salvage a
poorly scripted interactive narrative consisting mostly of swordplay or
where the outcome to the narrative is inevitably the same, no matter
what paths you take or choices you make as you work through it.
This, surely, is not the secondary orality with which Ong concludes
Orality and Literacy, a superficial category that ignores the script
lurking behind every exchange of words on television and radio.

Instead, in the mid-twentieth century we entered a world increasingly dominated by scripted orality on radio and television, in films and narratives like *The Last Express.* Or, in the case of interactive narratives, scripting squared, because every interactive text requires a script—or subscript—that anticipates the potential interests and desires of its readers, their possible moves and actions carefully plotted and blocked, choreographed as the foundation for the script, for the scenes, lines, and sequences readers encounter.

As oxymorons go, scripted orality is a fitting label for media and genres themselves rich in paradox. For realism that becomes more real the more it is manufactured. For stories that require more writing than print novels in the form of dialogue, narrative scripts, and scripted interactions that tether story firmly to a spectrum of potential readerly interactions with the text. For genres that seem at once to put readers closer to the action in stories, freeing them to explore realistically scenarios and settings long vanished, irrecoverable. Genres that can also physically reify and make palpable their authors' intentions even as they offer their readers a freedom impossible in print.

This future may be closer than even the techno-enthusiasts imagine. One of the novels shortlisted for the 1998 Booker Prize—arguably the United Kingdom's most notable literary award—was *The Angels of Russia* by Patricia Le Roy, published by the firm Online Originals, which offers thirty-three titles over the Internet, available for downloading to your PC or PalmPilot for the princely sum of roughly seven dollars. While acknowledging that *Angels of Russia* did not exactly match the definition of "book" listed in the Booker Prize rules that more or less restrict books to pages sandwiched between covers, Booker administrator Martyn Goff reviewed the title, which eventually made its way onto the Booker shortlist—on pages printed on paper, couriered over from Internet publisher and delivered by hand.[27]

The book is dead. Long live the book—whatever its form.

Notes

Chapter 1

1. Anthony J. Niez and Norman N. Holland, "Interactive Fiction," *Critical Inquiry* 11 (1984): 111.

2. Articles on readers and hypertexts include Stuart Moulthrop and Nancy Kaplan, "Something to Imagine: Literature, Composition, and Interactive Fiction," *Computers and Composition* 9, no. 1 (1991): 7–24; Moulthrop and Kaplan, "They Became What They Beheld: The Futility of Resistance in the Space of Hypertext Writing," in *Literacy and Computers: The Complications of Teaching and Learning with Technology*, ed. Susan Hilligoss and Cynthia L. Selfe (New York: Modern Language Association, 1991), 105–32; Michael Joyce, "Siren Shapes: Exploratory and Constructive Hypertexts," *Academic Computing* 3, no. 4 (1988): 10–14, 37–42; J. Yellowlees Douglas, "Plucked from the Labyrinth: Intention, Interpretation, and Interactive Narratives," in *Knowledge in the Making: Challenging the Text in the Classroom*, ed. Bill Corcoran, Mike Hayhoe, and Gordon M. Pradl (Portsmouth, N.H.: Boynton/Cook, 1994), 179–92; Douglas, "Gaps, Maps, and Perception: What Hypertext Readers (Don't) Do," *Perforations* 3 (spring–summer 1992): 1–13.

3. Jurgen Fauth, "Poles in Your Face: The Promises and Pitfalls of Hyperfiction," *Mississippi Review* 2, no. 6 (September 1995): <http://orca.st.usm.edu/mrw/backweb.html>.

4. Thomas Swiss, "Music and Noise: Marketing Hypertexts," review of Eastgate Systems, Inc., *Post Modern Culture* 7, no. 1 (1996): <http://jefferson.village.virginia.edu/pmc/text-only/issue.996/review-4.996>.

5. Espen J. Aarseth, *Cybertext: Perspectives on Ergodic Literature* (Baltimore: Johns Hopkins University Press, 1997), 49.

6. Aarseth, *Cybertext*, 3.

7. See Thomas Etter and William Chamberlain, *Racter* (Northbrook, Ill.: Mindscape, 1984); James Richard Meehan, "The Metanovel: Writing Stories by Computer," Ph.D. diss., Yale University, 1976.

8. Aarseth, *Cybertext*, 49.

9. Aarseth, *Cybertext*, 27.

10. Janet H. Murray, *Hamlet on the Holodeck: The Future of Narrative in Cyberspace* (New York: Free Press, 1997), 74.

11. Murray defines digital environments as procedural, participatory (these first two terms making "up most of what we mean by the vaguely used word interactive")—spatial, and encyclopedic, making them both explorable and extensive (*Hamlet on the Holodeck*, 71).

12. Jay David Bolter, *Writing Space: The Computer, Hypertext, and the History of Writing* (Hillsdale, N.J.: Lawrence Erlbaum Associates, 1991), 122.

13. Bolter, *Writing Space*, 121.

14. I use the term *narrative* here as it is most commonly used, to mean an account or story, as distinct from its more specific use as a structuralist term denoting the way and means by which the plot of a text is communicated. Some critics, most notably Aarseth, betray confusion with the term, despite his reasonably sophisticated discussion of narratology in *Cybertext*, most notably on p. 85, where he confuses narration and narrative.

15. Sven Birkerts, *The Gutenberg Elegies: The Fate of Reading in an Electronic Age* (Boston: Faber and Faber, 1994), 158.

16. Robert Coover, "The End of Books," *New York Times Book Review*, June 21, 1992, 1; emphasis added.

17. Laura Miller, "www.claptrap.com," *New York Times Book Review*, March 15 1998, 43.

18. Coover, "The End of Books," 25.

19. See, for example, Wen Stephenson, "Over the Edge: A Postmodern Freefall into Cyberspace," *Atlantic Unbound*, December 4, 1997.

20. Elizabeth L. Eisenstein, *The Printing Revolution in Early Modern Europe* (London: Cambridge University Press, 1983), 8.

21. Wolfgang Iser, *The Act of Reading: A Theory of Aesthetic Response* (Baltimore: Johns Hopkins University Press, 1978), xi.

22. For a discussion on differences between writing for disk and for the World Wide Web, see the interview with Michael Joyce in "The End of the Story," interview by Ralph Lombreglia, *Atlantic Unbound*, November 1996.

Chapter 2

1. Douglas Adams, introduction to *Douglas Adams' Starship Titanic: The Official Strategy Guide*, by Neil Richards (New York: Three Rivers Press, 1998), 8.

2. Review of *Douglas Adams' Starship Titanic*, *Kirkus Reviews*, September 1, 1997.

3. The analysis was conducted by metallurgist Timothy Foecke, as part of a team of marine forensic experts investigating the disaster headed by

naval architect William Garzke. See "For Want of Rivets, *Titanic* Was Lost, Scholars Speculate," *Boston Globe,* January 28, 1998, A10.

4. I rely on *readers,* rather than the more customary *users* or *players,* when referring to digital narratives, as these texts require readings of letters, diaries, menus, and other written materials used in their plots, just as they also require those navigating through them to "read" or interpret characters' motivations and intentions—a far cry from the usual ax swinging and gunslinging required of users in interactive games.

5. Victor Nell, *Lost in a Book: The Psychology of Reading for Pleasure* (New Haven: Yale University Press, 1988), 2.

6. Stephen Manes, review of *Riven, New York Times,* November 4, 1997.

7. Joyce, conversation with author, November 1986. Jay David Bolter, Michael Joyce, and John B. Smith are the creators of Storyspace hypertext authoring software for Macintosh and Windows (Cambridge, Mass.: Eastgate Systems, 1991).

8. Jorge Luis Borges, "An Autobiographical Essay," in *The Aleph and Other Stories: 1933–1969,* trans. Norman Thomas Di Giovanni (New York: Dutton, 1970), 249–50.

9. Eisenstein, *Printing Revolution,* 84.

10. For a survey of different definitions of hypertext, see J. Yellowlees Douglas, "Social Impacts of Computing: The Framing of Hypertext—Revolutionary for Whom?" *Social Science Computer Review* 11, no. 4 (1993): 417–28.

11. Murray, *Hamlet on the Holodeck,* 212.

12. Michael Heim, *The Metaphysics of Virtual Reality* (New York: Oxford University Press, 1993), 30.

13. Richard Howard, foreword to *The Pleasure of the Text,* by Roland Barthes, trans. Richard Miller (New York: Hill and Wang, 1975), vii.

14. For the most thorough and comprehensive example of this, see George Landow, *Hypertext: The Convergence of Technology and Literary Theory* (Baltimore: Johns Hopkins University Press, 1992).

15. Jacques Derrida, *Of Grammatology,* trans. Gayatri Chakravorty Spivak (Baltimore: Johns Hopkins University Press, 1976), 84.

16. Paul Saenger, "The Separation of Words and the Physiology of Reading," in *Literacy and Orality,* ed. David R. Olson and Nancy Torrance (Cambridge: Cambridge University Press, 1991), 209.

17. Richard Ziegfeld, "Interactive Fiction: A New Literary Genre?" *New Literary History* 20 (1989): 358.

18. See, for example, J. Yellowlees Douglas, "Wandering through the Labyrinth: Encountering Interactive Fiction," *Computers and Composition* 6, no. 3 (1989): 93–103; and Bolter, *Writing Space.*

19. Plato, *Phaedrus,* trans. Walter Hamilton (London: Penguin, 1973), 97.

20. In this particular instance, an early version of *afternoon* found its way into my hands by that reliable analog version of the information super-highway, Sneaker Net. I was given a copy of Michael Joyce's hypertext along with a beta test copy of Storyspace, and tracked the living, breathing author down through the White Pages.

21. Vannevar Bush, "As We May Think," *Atlantic* 176, no. 1 (1945): 101–2.

22. James M. Nyce and Paul Kahn, "A Machine for the Mind: Vannevar Bush's Memex," in *From Memex to Hypertext: Vannevar Bush and the Mind's Machine,* ed. James M. Nyce and Paul Kahn (New York: Academic Press, 1992), 63.

23. Bush, "As We May Think," 105.

24. Jean-Paul Sartre, *What Is Literature?* trans. Bernard Frechtman (London: Philosophical Library, 1949).

25. Robert C. Holub, *Reception Theory: A Critical Introduction* (London: Methuen, 1984), 24–25.

26. Iser, *The Act of Reading,* 101.

27. Sartre, *What Is Literature?* 24.

28. Iser, *The Act of Reading,* 66.

29. This passage was used in tests conducted by psychologists in testing reading comprehension. See Paul T. Wilson and Richard C. Anderson, "What They Don't Know Will Hurt Them: The Role of Prior Knowledge in Comprehension," in *Reading Comprehension: From Research to Practice,* ed. Judith Orasanu (Hillsdale, N.J.: Lawrence Erlbaum Associates, 1986), 34.

30. Shlomith Rimmon-Kenan, *Narrative Fiction: Contemporary Poetics* (New York: Methuen, 1983), 121.

31. The example is modeled after experiments in Alison Black, Paul Freeman, and P. N. Johnson-Laird, "Plausibility and the Comprehension of Text," *British Journal of Psychology* 77 (1986): 58.

32. See Deidre Sperber and David Wilson, "Mutual Knowledge and Relevance in Theories of Comprehension," in *Mutual Knowledge,* ed. N. V. Smith (London: Academic Press, 1982), 42–59; Black, Freeman, and Johnson-Laird, "Plausibility," 51–62.

Chapter 3

1. Walter J. Ong, *Orality and Literacy: The Technologizing of the Word* (London: Methuen, 1982), 102–3.

2. Jay David Bolter, "Topographic Writing: Hypertext and the Electronic Writing Space," in *Hypermedia and Literary Studies,* ed. Paul Delany and George P. Landow (Cambridge: MIT Press, 1991), 105.

3. George P. Landow, "Hypertext, Hypermedia, and Literary Studies: The State of the Art," in Delany and Landow, *Hypermedia and Literary Studies,* 3.

4. Theodor Holm Nelson, *Literary Machines* (Bellevue, Wash.: OWL Systems, 1987), 1.

5. John M. Slatin, "Reading Hypertext: Order and Coherence in a New Medium," *College English*, 52, no. 8 (1990): 876. More recent definitions of hypertext also emphasize the medium's multilinear and multisequential aspects. See Landow, *Hypertext*, 4.

6. I use the term *medium* in a way analogous to its use in painting—where artists work with oil, pastels, watercolors, or etchings, the different tools causing them to render vastly different effects. At the same time, painting generally is also a medium, as the term is used to refer to media like radio, television, and film. Similarly, as used here, the medium for hypertext fiction is different from that used in creating and reading digital narratives—and digital environments are also the medium in which writers like Joyce and Moulthrop create interactive narratives.

7. Murray, *Hamlet on the Holodeck*, 133.

8. See Slatin, "Reading Hypertext," 871. See also Frank Smith, *Understanding Reading: A Psycholinguistic Analysis of Reading and Learning to Read*, 3d ed. (New York: Holt, Rinehart and Winston, 1982), 76–77.

9. Slatin, "Reading Hypertext," 871.

10. Bolter discusses tables of contents as print examples of hierarchical maps in *Writing Space*, 22.

11. Bolter, *Writing Space*, 124.

12. Student Andrew Sussman, quoted in Moulthrop and Kaplan, "Something to Imagine," 16.

13. Tom Trabasso, Tom Secco, and Paul Van Den Broek, "Causal Cohesion and Story Coherence," in *Learning and Comprehension of Text*, ed. Heinz Mandl, Nancy L. Stein, and Tom Trabasso (Hillsdale, N.J.: Lawrence Erlbaum Associates, 1984), 87.

14. Ziegfeld, "Interactive Fiction," 363.

15. Barthes, "Death of the Author," in *Image-Music-Text*, trans. Stephen Heath (New York: Hill and Wang, 1977), 148.

16. Lippman is quoted in Stewart Brand, *The Media Lab: Inventing the Future at MIT* (New York: Viking, 1987), 46–49.

17. I discuss the Mood Bar™ as interface more fully in "Virtual Intimacy™ and the Male Gaze Cubed: Interacting with Narratives on CD-ROM," *Leonardo* 29, no. 3 (1996): 207–13.

18. Aarseth, *Cybertext*, 49.

19. Murray argues, "If we ask the interactor to pick from a menu of things to say, we limit agency and remind them of the fourth wall. Some CD-ROM stories give the interactor the task of deciding the mood or tone of a spoken response rather than picking a statement from a list of possible things to say. This is a more promising route because it seems less mechanical, although the mood selector is often a menu on a slider bar that is outside the story" (*Hamlet on the Holodeck*, 190–91).

20. *Titanic*'s narratives suggest that you have jumped tracks by, for

example, barring access to the First Class lounge or Café Parisien through your steward's telling you repeatedly that both rooms are closed and showing you the door. Narratives like *Gadget* bar your exit from the story's railway stations and museums until you milk all the requisite clues from each scene. Others, such as *Midnight Stranger,* shuttle the reader into empty comedy clubs and restaurants where there is nobody to interact with, while narratives like *Who Killed Taylor French?* hustle you along with gentle prods from your assistant, chewings out by a police superior, and a loudly ticking clock signifying how little time remains before you need to swear out a warrant for your suspect's arrest.

21. In *The Last Express,* discussed at length in chapter 7, however, there is no "dead zone": anywhere you wander on the train at any time provides opportunities for interactions with other characters. If, for example, you do not choose to sleep during the small hours, you can still interact with characters on board the Orient Express—if they are also awake—and suffer the repercussions of sleep deprivation the following day.

22. Michael Joyce, "Selfish Interactions: Subversive Texts and the Multiple Novel," in *Of Two Minds: Hypertext Pedagogy and Poetics* (Ann Arbor: University of Michigan Press, 1995), 144.

23. B. K. Britton, A. Piha, J. Davis, and E. Wehausen, "Reading and Cognitive Capacity Usage: Adjunct Question Effects," *Memory and Cognition* 6 (1978): 266–73.

24. Bolter, *Writing Space,* 144.

25. Stuart Moulthrop, "Reading from the Map: Metonymy and Metaphor in the Fiction of Forking Paths," in Delany and Landow, *Hypermedia and Literary Studies,* 127.

26. Peter Brooks, *Reading for the Plot: Design and Intention in Narrative* (New York: Vintage, 1985), 23.

27. Moulthrop, "Reading from the Map," 128.

28. Bolter, *Writing Space,* 143.

29. Joseph Frank, "Spatial Form in Modern Literature," in *Essentials of the Theory of Fiction,* ed. Michael Hoffman and Patrick Murphy (Durham: Duke University Press, 1988), 85–100.

30. See W. J. T. Mitchell, "Spatial Form in Literature: Toward a General Theory," in *The Language of Images,* ed. W. J. T. Mitchell (Chicago: University of Chicago Press, 1980), 284. See also Jeffrey R. Smitten, "Spatial Form and Narrative Theory," in *Spatial Form in Narrative,* ed. Jeffrey R. Smitten and Ann Daghistany (Ithaca: Cornell University Press, 1981), 19–20.

31. Richard Lanham, "The Electronic Word: Literary Study and the Digital Revolution," *New Literary History* 20 (1989): 269.

32. Bolter, *Writing Space,* 124–25.

33. Ziegfeld, "Interactive Fiction," 352.

34. Slatin, "Reading Hypertext," 872.

35. Lawrence Durrell, *The Alexandria Quartet* (London: Faber and Faber, 1962), 249.

36. Durrell, *The Alexandria Quartet*, 555.

37. Barthes, *Pleasure of the Text*, 9.

38. Michael Joyce, *afternoon, a story* (Cambridge, Mass.: Eastgate Systems, 1990), "asks."

39. Coover, "The Babysitter," in *Pricksongs and Descants* (New York: Plume, 1969), 239.

40. Jorge Luis Borges, "The Garden of Forking Paths," in *Fictions*, trans. Anthony Kerrigan (London: John Calder, 1985), 91.

41. Moulthrop, "Reading from the Map," 124.

42. Bolter, *Writing Space*, 143.

Chapter 4

1. Louis Gianetti, *Understanding Movies*, 5th ed. (Englewood Cliffs, N.J.: Prentice-Hall, 1990), 113.

2. James Monaco, *How to Read a Film: The Art, Technology, Language, History, and Theory of Film and Media* (New York: Oxford University Press, 1981), 322–23.

3. Jerome Bruner, *Actual Minds, Possible Worlds* (Cambridge: Harvard University Press, 1986), 17.

4. A. Michotte, *The Perception of Causality* (New York: Basic Books, 1963), 375.

5. Fritz Heider and Marianne Simmel, "An Experimental Study of Apparent Behavior," *American Journal of Psychology* 57 (1944): 23–41.

6. Bruner, *Actual Minds, Possible Worlds*, 18.

7. Alan M. Leslie and Stephanie Keeble, "Do Six-Month-Old Infants Perceive Causality?" *Cognition* 25 (1987): 265–88.

8. Jeremy Campbell, *Winston Churchill's Afternoon Nap: A Wide-Awake Inquiry into the Human Nature of Time* (London: Paladin, 1989), 267–68.

9. See, for example, John Black and Gordon H. Bower, "Episodes as Chunks in Narrative Memory," *Journal of Verbal Learning and Verbal Behavior* 18 (1979): 309–18.

10. Campbell, *Winston Churchill's Afternoon Nap*, 269. See also Roger C. Schank, *Dynamic Memory: A Theory of Reminding and Learning in Computers and People* (Cambridge: Cambridge University Press, 1982), 14.

11. Monaco, *How to Read Film*, 73.

12. Slatin, "Reading Hypertext," 877.

13. Using Storyspace or Javascript "cookies," however, writers can attach conditions to links, requiring readers to have visited specific screens or segments of text before they can even see some navigational options.

While these scripts or Boolean strings do contain text of a sort, they generally are not intended to be seen by readers.

14. Rimmon-Kenan, *Narrative Fiction,* 127.

15. Wolfgang Iser, "The Reading Process: A Phenomenological Approach," in *Reader-Response Criticism: From Formalism to Post-Structuralism,* ed. Jane P. Tompkins (Baltimore: Johns Hopkins University Press, 1980), 285.

16. Laurence Sterne, *The Life and Opinions of Tristram Shandy,* 1759 (London: Penguin Books, 1967), 127.

17. Bolter, *Writing Space,* 140.

18. Bolter, *Writing Space,* 141.

19. Teun van Dijk, *Macrostructures: An Interdisciplinary Study of Global Structures in Discourse, Interaction, and Cognition* (Hillsdale, N.J.: Lawrence Erlbaum Associates, 1980), 9–11.

20. Roger C. Schank and Robert P. Abelson, *Scripts, Plans, Goals and Understanding: An Inquiry into Human Knowledge Structures* (Hillsdale, N.J.: Lawrence Erlbaum Associates, 1977), 37–39.

21. Bolter, *Writing Space,* 143.

22. Slatin, "Reading Hypertext," 872.

23. Moulthrop, "Reading from the Map," 128.

24. Brooks, *Reading for the Plot,* 93–94.

25. Bruner, *Actual Minds, Possible Worlds,* 36.

26. Frank Kermode, "Secrets and Narrative Sequence," in *On Narrative,* ed. W. J. T. Mitchell (Chicago: University of Chicago Press, 1981), 88.

27. Brooks, *Reading for the Plot,* 94.

28. Borges, "Garden of Forking Paths," 89.

29. Borges, "Garden of Forking Paths," 92.

30. Trabasso, Secco, and Van Den Broek, "Causal Cohesion," 87; Barbara Herrnstein-Smith, *Poetic Closure: A Study of How Poems End* (Chicago: University of Chicago Press, 1987), 36.

31. Moulthrop and Kaplan, "Something to Imagine," 16.

32. David Riesman, *The Lonely Crowd: A Study of the Changing American Character* (New Haven: Yale University Press, 1950), 259.

Chapter 5

1. Brooks, *Reading for the Plot,* 20.

2. Brooks, *Reading for the Plot,* 22.

3. Aristotle *Poetics,* trans. S. H. Butcher, in *Critical Theory since Plato,* ed. Hazard Adams (New York: Harcourt Brace Jovanovich, 1971), 52.

4. Frank Kermode, *The Sense of an Ending: Studies in the Theory of Narrative Fiction* (New York: Oxford University Press, 1966), 7, 17.

5. Kermode, *Sense of an Ending,* 17.

6. Smith, *Understanding Reading,* 77.

7. Brooks, *Reading for the Plot*, 23.

8. Brooks, *Reading for the Plot*, 52.

9. Kermode, *Sense of an Ending*, 19.

10. Herrnstein-Smith, *Poetic Closure*, 30.

11. Rick Barba, *"The Last Express": The Official Strategy Guide* (Rocklin, Calif.: Prima Publishing, 1997), 1, hereafter cited as *Official Strategy Guide*.

12. Barba, *Official Strategy Guide*, 1.

13. Bolter, *Writing Space*, 144.

14. Ziegfeld, "Interactive Fiction," 363.

15. Michael Joyce, conversation with the author, October 1991.

16. Brooks, *Reading for the Plot*, 280–81.

17. Joyce, *afternoon*, "1/".

18. Joyce, *afternoon*, "2/".

19. Herrnstein-Smith, *Poetic Closure*, 36.

20. Michael Joyce, conversation with the author, October 1991.

21. Stuart Moulthrop, "Hypertext and 'the Hyperreal,'" *Hypertext '89 Proceedings* (Pittsburgh: Association for Computing Machinery, 1989), 262.

22. Frank, "Spatial Form," 85.

23. David Mickelsen, "Types of Spatial Structure in Narrative," in Smitten and Daghistany, *Spatial Form in Narrative*, 74.

24. Joseph Kestner, "The Novel and the Spatial Arts," in Smitten and Daghistany, *Spatial Form in Narrative*, 128.

25. Ivo Vidan, "Time Sequence in Spatial Fiction," in Smitten and Daghistany, *Spatial Form in Narrative*, 133.

26. Frank, "Spatial Form," 88.

27. Mitchell, "Spatial Form in Literature," 272.

28. See, for example, Rimmon-Kenan, *Narrative Fiction*, 121.

29. Stanley Fish, *Is There a Text in This Class? The Authority of Interpretive Communities* (Cambridge: Harvard University Press, 1980), 318.

30. Fish, *Is There a Text*, 327.

31. Bolter, *Writing Space*, 136.

32. Jay David Bolter, "The Shapes of *WOE*," *Writing on the Edge* 2, no. 2 (1991): 91.

33. Johndan Johnson-Eilola, "'Trying to See the Garden': Interdisciplinary Perspectives on Hypertext Use in Composition Instruction," *Writing on the Edge* 2, no. 2 (1991): 104–5.

34. Umberto Eco, *The Open Work*, trans. Anna Cancogni (Cambridge: Harvard University Press, 1989), 15–16.

35. Brooks, *Reading for the Plot*, 23.

Chapter 6

1. Eco, *The Open Work*, 3.

2. Eco, *The Open Work*, 11.

3. Eco, *The Open Work*, 15.

4. W. K. Wimsatt, and Monroe C. Beardsley, "The Intentional Fallacy," in *The Verbal Icon* (Lexington: University of Kentucky Press, 1954).

5. Lionel Trilling, "Freud and Literature," in *The Liberal Imagination: Essays on Literature and Society* (New York: Viking, 1968).

6. Eco, *The Open Work*, 99.

7. Joyce, *afternoon*, "Hidden Wren."

8. Eco, *The Open Work*, 15.

9. Steve Rosenthal, "Douglas Engelbart," *Electric Word* 18 (1990): 21.

10. Eco, *The Open Work*, 102.

11. See J. Yellowlees Douglas, "'Nature' versus 'Nurture': The Three Paradoxes of Hypertext," *Readerly/Writerly Texts* 3, no. 2 (1996): 185–207.

Chapter 7

1. Miller, "www.claptrap.com," 43.

2. Birkerts, *The Gutenberg Elegies*, 163.

3. Britton et al., "Reading and Cognitive Capacity."

4. Q. D. Leavis, *Fiction and the Reading Public* (London: Chatto and Windus, 1932), 50.

5. Marie-Laure Ryan, *Possible Worlds, Artificial Intelligence, and Narrative Theory* (Bloomington: Indiana University Press, 1991), 150.

6. Bruner, *Actual Minds, Possible Worlds*, 17.

7. Hayden White, "The Value of Narrativity in the Representation of Reality," in Mitchell, *On Narrative*, 23.

8. R. D. Altick, *The English Common Reader: A Social History of the Mass Reading Public, 1800–1900* (Chicago: University of Chicago Press, 1957), 28–29.

9. Robert de Beaugrande and Benjamin N. Colby, "Narrative Models of Action and Interaction," *Cognitive Science* 3 (1979): 45.

10. Beaugrande and Colby, "Narrative Models of Action," 49.

11. Beaugrande and Colby, "Narrative Models of Action," 50.

12. John G. Cawelti, *Adventure, Mystery, and Romance: Formula Stories as Art and Popular Culture* (Chicago: University of Chicago Press, 1976), 15.

13. Beaugrande and Colby, "Narrative Models of Action," 50.

14. Bert O. States, *Dreaming and Storytelling* (Ithaca: Cornell University Press, 1993), 112.

15. This feature is unique to Joyce's fiction and not to Web-based fiction generally, as both "Trip" and *253* feature micronarratives that exist at the level of one or two segments of text that are part of a larger macronarrative in each hypertext: Will the narrator ever track down his ex-girlfriend and return her two kids to her? What will become of the train hurtling toward Elephant and Castle—and will anyone survive the crash ahead?

16. Readers can, however, also navigate by means of a single text link on each screen that they view, although, once they choose the link, it vanishes.

17. See Greg Ulmer's "A Response to *Twelve Blue* by Michael Joyce," *Post Modern Culture* 8, no. 1 (1997): <http://jefferson.village.virginia/edu/pmc/issue.997/ulmer.997.html>.

18. This label was first suggested by a Vassar student of Joyce's, Josh Lechner, during a lecture I gave at Vassar in May 1994; Joyce takes up Lechner's label once more in "Ordinary Fiction," *Paradoxa* 4, no. 11 (spring 1999): 510–26.

19. Joyce, "Ordinary Fiction," 11–12.

20. Tom Wolfe, *The New Journalism* (London: Picador, 1972), 46–56.

21. E. M. Forster, *Aspects of the Novel* (New York: Harcourt Brace, 1927).

22. For details on the painstaking recreation of the Orient Express as it appeared on the brink of World War I, see "The Making of T*he Last Express:* An Interview with Jordan Mechner and Tomi Pierce," in Barba, *Official Strategy Guide;* Charles Egan, "Render on the Orient Express," *New Media,* March 24, 1998, 25.

23. Only half the conversations, actions, and interactions readers can enjoy in *The Last Express* are relevant to either its macro- or its attendant microplots. See Barba, *Official Strategy Guide,* 1.

24. Murray, *Hamlet on the Holodeck,* 181.

25. You can, of course, be economical even with surplus: Joyce conveys ordinarymindedness in only ninety-six segments of text in "Twelve Blue." Similarly, Rebecca Norton, Sophie de Bretheuil, Mahmud Makhta, and the four-woman harem he chaperones do not even play incidental roles in either macroplot or the microplots it entails in *The Last Express.* The harem and its escort, however, flesh out the narrative with minimal appearances at doors of compartments, flitting shadowlike in the compartment carriageway as François plays, giggling and gossiping quietly behind closed doors.

26. Wolfe, *The New Journalism,* 49.

27. Sue Kelley, "A Literary Prize Gets Wired," *New Yorker,* September 14, 1998, 30–31.

Bibliography

Aarseth, Espen J. *Cybertext: Perspectives on Ergodic Literature.* Baltimore: Johns Hopkins University Press, 1997.

Adams, Douglas. *Douglas Adams' Starship Titanic.* CD-ROM for Windows. Simon and Schuster Interactive, 1998.

Allinson, Lesley, and Nick Hammond. "A Learning Support Environment: The Hitch-hiker's Guide." In *Hypertext: Theory into Practice,* ed. Ray McAleese. Oxford: Intellect Books, 1990.

Althusser, Louis. *The Future Lasts Forever: A Memoir.* Trans. Richard Veasey and ed. Olivier Corpet and Yann Moulier Boutang. New York: New Press, 1993.

Altick, R. D. *The English Common Reader: A Social History of the Mass Reading Public, 1800–1900.* Chicago: University of Chicago Press, 1957.

Aristotle. *Poetics.* Trans. S. H. Butcher. In *Critical Theory since Plato,* ed. Hazard Adams. New York: Harcourt Brace Jovanovich, 1971.

Arnold, Mary-Kim. "Lust." *Eastgate Quarterly Review of Hypertext* 1, no. 2 (1993). Storyspace software for Macintosh and Windows.

Barnes, Julian. *Flaubert's Parrot.* New York: McGraw-Hill, 1984.

Barth, John. "The Literature of Exhaustion." *Atlantic* 220, no. 2 (1967): 29–34.

———. "Lost in the Funhouse." In *Lost in the Funhouse: Fiction for Print, Tape, Live Voice.* Garden City, N.Y.: Doubleday, 1968.

———. "The State of the Art." *Wilson Quarterly* (spring 1996): 36–45.

Barthes, Roland. *The Pleasure of the Text.* Trans. Richard Miller. New York: Hill and Wang, 1975.

———. "The Death of the Author." In *Image, Music, Text,* trans. Stephen Heath. New York: Hill and Wang, 1977.

———. "From Work to Text." In *Textual Strategies: Perspectives in Post-Structural Criticism,* ed. Josué Harari. Ithaca: Cornell University Press, 1979.

Beaugrande, Robert de, and Benjamin N. Colby. "Narrative Models of Action and Interaction." *Cognitive Science* 3 (1979): 43–66.

Benjamin, Walter. "The Storyteller." In *Illuminations,* trans. Harry Zohn and ed. Hannah Arendt. New York: Schocken, 1969.

Bernstein, Mark. "The Bookmark and the Compass: Orientation Tools for Hypertext Users." *ACM SIGOIS Bulletin,* October 1988, 34–45.

———. "The Navigation Problem Reconsidered." In *Hypertext/Hypermedia Handbook,* ed. Emily Berk and James Devlin. New York: McGraw-Hill, 1991.

Bernstein, Mark, and Erin Sweeney. *The Election of 1912.* Storyspace software for Macintosh. Cambridge, Mass.: Eastgate Systems, 1989.

Bernstein, Mark, Jay David Bolter, Michael Joyce, and Elli Mylonas. "Architectures for Volatile Hypertext." In *Hypertext '91 Proceedings.* San Antonio: Association for Computing Machinery, 1991.

Birkerts, Sven. *The Gutenberg Elegies: The Fate of Reading in an Electronic Age.* Boston: Faber and Faber, 1994.

Black, Alison, Paul Freeman, and P. N. Johnson-Laird. "Plausibility and the Comprehension of Text." *British Journal of Psychology* 77 (1986): 51–62.

Black, John, and Gordon H. Bower. "Episodes as Chunks in Narrative Memory." *Journal of Verbal Learning and Verbal Behavior* 18 (1979): 309–18.

Blackburn, Julia. *Daisy Bates in the Desert.* New York: Vintage, 1995.

Bolter, Jay David. "The Shapes of *WOE.*" *Writing on the Edge* 2, no. 2 (1991): 90–91.

———. "Topographic Writing: Hypertext and the Electronic Writing Space." In *Hypermedia and Literary Studies,* ed. Paul Delany and George P. Landow. Cambridge: MIT Press, 1991.

———. *Writing Space: A Hypertext.* Storyspace software for Macintosh. Cambridge, Mass.: Eastgate Systems, 1990.

———. *Writing Space: The Computer, Hypertext and the History of Writing.* Hillsdale, N.J.: Lawrence Erlbaum Associates, 1991.

Borges, Jorge Luis. "An Autobiographical Essay." In *The Aleph and Other Stories: 1933–1969,* trans. Norman Thomas Di Giovanni. New York: Dutton, 1970.

———. "The Book of Sand." In *The Book of Sand,* trans. Norman Thomas Di Giovanni. London: Penguin, 1980.

———. "The Garden of Forking Paths." In *Fictions,* trans. Anthony Kerrigan. London: John Calder, 1985.

Brand, Stewart. *The Media Lab: Inventing the Future at MIT.* New York: Viking, 1987.

Britton, B. K., A. Piha, J. Davis, and E. Wehausen. "Reading and Cognitive Capacity Usage: Adjunct Question Effects." *Memory and Cognition* 6 (1978): 266–73.

Brooks, Cleanth. *The Well Wrought Urn.* New York: Harcourt Brace Jovanovich, 1947.

Brooks, Peter. *Reading for the Plot: Design and Intention in Narrative.* New York: Vintage, 1985.

Brown, P. J. "Turning Ideas into Products: The Guide System." In *Proceedings of Hypertext '87.* Chapel Hill: University of North Carolina Press, 1987.

Bruner, Jerome. *Acts of Meaning.* Cambridge: Harvard University Press, 1990.

———. *Actual Minds, Possible Worlds.* Cambridge: Harvard University Press, 1986.

Bush, Vannevar. "As We May Think." *Atlantic* 176, no. 1 (1945): 101–8.

———. "The Inscrutable 'Thirties: Reflections upon a Preposterous Decade." In *From Memex to Hypertext: Vannevar Bush and the Mind's Machine,* ed. James M. Nyce and Paul Kahn. New York: Academic Press, 1992.

Campbell, Jeremy. *Winston Churchill's Afternoon Nap: A Wide-Awake Inquiry into the Human Nature of Time.* London: Paladin, 1989.

Cawelti, John G. *Adventure, Mystery, and Romance: Formula Stories as Art and Popular Culture.* Chicago: University of Chicago Press, 1976.

Chatman, Seymour. *Story and Discourse: Narrative Structure in Fiction and Film.* Ithaca: Cornell University Press, 1978.

———. "What Novels Can Do That Films Can't (and Vice Versa)." In *On Narrative,* ed. W. J. T. Mitchell. Chicago: University of Chicago Press, 1981.

Colomb, Gregory G. "Cultural Literacy and the Theory of Meaning: Or What Educational Theorists Need to Know about How We Read." *New Literary History* 20 (1989): 411–49.

Conklin, Jeffrey. "A Survey of Hypertext." *IEEE Computer* 20 (September 1987): 15–34.

Conrad, Joseph. "Henry James: An Appreciation." In *The Collected Works of Joseph Conrad.* London: J. M. Dent, 1905.

Coover, Robert. "The End of Books." *New York Times Book Review,* June 21, 1992, 1, 23–24.

———. "Hyperfiction: Novels for the Computer." *New York Times Book Review,* August 29, 1993, 1, 8–12.

———. *Pricksongs and Descants.* New York: Plume, 1969.

Cortázar, Julio. *Hopscotch.* Trans. Gregory Rabassa. New York: Random House, 1966.

Crane, Gregory. "Composing Culture: The Authority of an Electronic Text." *Current Anthropology* 32, no. 3 (1991): 293–311.

Davenport, Elisabeth, and Blaise Cronin. "Hypertext and the Conduct of Science." *Journal of Documentation* 46, no. 3 (1990): 175–92.

Davis, R., and G. de Jong. "Prediction and Substantiation: Two Processes That Comprise Understanding." *Proceedings of the International Joint Conference on Artificial Intelligence* 5 (1979): 217–22.

Demos, John. *The Unredeemed Captive: A Family Story from Early America.* New York: Vintage, 1994.

Derrida, Jacques. *Of Grammatology.* Trans. Gayatri Chakravorty Spivak. Baltimore: Johns Hopkins University Press, 1976.

———. "Structure, Sign, and Play in the Discourse of the Human Sciences." In *The Structuralist Controversy: The Languages of Criticism in the Sci-*

ences of Man, ed. Richard S. Macksey and Eugenio Donato. Baltimore: Johns Hopkins University Press, 1972.

Douglas, J. Yellowlees. "I Have Said Nothing." *Eastgate Quarterly Review of Hypertext* 1, no. 2 (1994). Storyspace for Macintosh and Windows.

———. "'Nature' versus 'Nurture': The Three Paradoxes of Hypertext." *Readerly/Writerly Texts* 3, no. 2 (1996): 185–207.

———. "Plucked from the Labyrinth: Intention, Interpretation, and Interactive Narratives." In *Knowledge in the Making: Challenging the Text in the Classroom,* ed. Bill Corcoran, Mike Hayhoe, and Gordon M. Pradl. Portsmouth, N.H.: Boynton/Cook, 1994.

———. "Social Impacts of Computing: The Framing of Hypertext—Revolutionary for Whom?" *Social Science Computer Review* 11, no. 4 (1993): 417–28.

———. "Understanding the Act of Reading: The *WOE* Beginner's Guide to Dissection." *Writing on the Edge* 2, no. 2 (1991): 112–26.

———. "Virtual Intimacy™ and the Male Gaze Cubed: Interacting with Narratives on CD-ROM." *Leonardo* 29, no. 3 (1996): 207–13.

———. "Wandering through the Labyrinth: Encountering Interactive Fiction." *Computers and Composition* 6, no. 3 (1989): 93–103.

———. "Where the Senses Become a Stage and Reading Is Direction: Performing the Texts of Virtual Reality and Interactive Fiction." *Drama Review* 37, no. 4 (1993): 18–37.

Durrell, Lawrence. *The Alexandria Quartet.* London: Faber and Faber, 1962.

Eco, Umberto. *The Open Work.* 1962. Trans. Anna Cancogni. Cambridge: Harvard University Press, 1989.

Edward, Deborah M., and Lynda Hardman. "'Lost in Hyperspace': Cognitive Mapping and Navigation in a Hypertext Environment." In *Hypertext: Theory into Practice,* ed. Ray McAleese. Oxford: Intellect Books, 1990.

Egan, Charles. "Render on the Orient Express." *New Media,* March 24, 1998, 25.

Eisenstein, Elizabeth L. *The Printing Press as an Agent of Change: Communications and Cultural Transformations in Early Modern Europe.* 2 vols. Cambridge: Cambridge University Press, 1979.

———. *The Printing Revolution in Early Modern Europe.* London: Cambridge University Press, 1983.

Eliot, T. S. *The Waste Land.* In *Selected Poems.* New York: Harcourt Brace Jovanovich, 1964.

Engelbart, Douglas. "A Conceptual Framework for the Augmentation of Man's Intellect." In *Computer-Supported Cooperative Work: A Book of Readings,* ed. Irene Greif. San Mateo: Morgan Kaufmann, 1988.

Etter, Thomas, and William Chamberlain. *Racter.* Northbrook, Ill.: Mindscape, 1984.

Fauth, Jurgen. "Poles in Your Face: The Promises and Pitfalls of Hyperfiction." *Mississippi Review* 2, no. 6 (September 1995): <http://orca.st.usm.edu/mrw/backweb.html>.

Febvre, Lucien, and Henri-Jean Martin. *The Coming of the Book: The Impact of Printing, 1450–1800.* Trans. David Gerard. London: Verso, 1990.

Finnegan, Ruth. *Literacy and Orality.* Oxford: Blackwell 1988.

Fish, Stanley. *Is There a Text in This Class? The Authority of Interpretive Communities.* Cambridge: Harvard University Press, 1980.

———. "Why No One's Afraid of Wolfgang Iser." In *Doing What Comes Naturally: Change, Rhetoric, and the Practice of Theory in Literary and Legal Studies.* Oxford: Clarendon Press, 1989.

Ford, Ford Madox. *The Good Soldier: A Tale of Passion.* 1915. Oxford: Oxford University Press, 1990.

Forster, E. M. *Aspects of the Novel.* New York: Harcourt Brace, 1927.

Foucault, Michel. *The Order of Things: An Archeology of the Human Sciences.* New York: Vintage, 1973.

———. "What Is an Author?" In *Language, Counter-Memory, Practice: Selected Essays and Interviews,* trans. Donald F. Bouchard and ed. Donald F. Bouchard and Sherry Simon. Ithaca: Cornell University Press, 1977.

Fowles, John. *The French Lieutenant's Woman.* New York: Signet, 1969.

Frank, Joseph. "Spatial Form in Modern Literature." In *Essentials of the Theory of Fiction,* ed. Michael Hoffman and Patrick Murphy. Durham: Duke University Press, 1988.

Freud, Sigmund. *Three Case Histories.* Trans. and ed. Philip Rieff. New York: Collier, 1963.

Garnham, Alan, Jane Oakhill, and P. N. Johnson-Laird. "Referential Continuity and the Coherence of Discourse." *Cognition* 11 (1982): 29–46.

Genette, Gérard. *Narrative Discourse: An Essay on Method.* Trans. Jane E. Lewin. Ithaca: Cornell University Press, 1970.

Geyh, Paula, Fred G. Leebron, and Andrew Levy, eds. *Postmodern American Fiction: A Norton Anthology.* New York: W. W. Norton, 1998.

Ghaoui, Claude, Steven M. George, Roy Rada, Martin Beer, and Janus Getta. "Text to Hypertext and Back Again." In *Computers and Writing: State of the Art,* ed. Noel Williams and Patrik Holt. Oxford: Intellect Books, 1992.

Gianetti, Louis. *Understanding Movies.* 5th ed. Englewood Cliffs, N.J.: Prentice Hall, 1990.

Gilligan, Shannon. *The Magic Death: Virtual Murder 2.* CD-ROM for Macintosh. Portland, Oreg.: Creative Multimedia Corporation, 1993.

———. *Who Killed Brett Penance? The Environmental Surfer: Murder Mystery 3.* CD-ROM for Macintosh. Portland, Oreg.: Creative Multimedia Corporation, 1995.

———. *Who Killed Sam Rupert? Virtual Murder 1.* CD-ROM for Macintosh. Portland, Oreg.: Creative Multimedia Corporation, 1993.

———. *Who Killed Taylor French? The Case of the Undressed Reporter:*

Murder Mystery 4. CD-ROM for Macintosh. Portland, Oreg.: Creative Multimedia Corporation, 1995.

Gombrich, E. H. *Art and Illusion: A Study in the Psychology of Pictorial Representation.* Princeton: Princeton University Press, 1956.

Goodwin, Simon, and Jeff Green. *Midnight Stranger.* CD-ROM for Macintosh. San Diego: Gazelle Technologies, 1994.

Goody, Jack. *The Domestication of the Savage Mind.* Cambridge: Cambridge University Press, 1977.

Guyer, Carolyn. *Quibbling.* Storyspace software for Macintosh and Windows. Cambridge, Mass.: Eastgate Systems, 1993.

Hagedorn, Stephen. "Technology and Economic Exploitation: The Serial as a Form of Narrative Presentation." *Wide Angle* 10, no. 4 (1986): 4–12.

Halasz, Frank. "Reflections on NoteCards: Seven Issues of the Next Generation of Hypermedia Systems." *Communications of the ACM* 31, no. 7 (1988): 836–52.

Hardison, O. B. *Disappearing through the Skylight: Culture and Technology in the Twentieth Century.* New York: Viking, 1989.

Havelock, Eric. *Preface to Plato.* Cambridge: Harvard University Press, 1963.

Hawke, David Freeman. *Nuts and Bolts of the Past: A History of American Technology, 1776–1860.* New York: Harper and Row, 1988.

Hayles, N. Katherine. "Corporeal Anxiety in Dictionary of the Khazars: What Books Talk about in the Late Age of Print When They Talk about Losing Their Bodies." *Modern Fiction Studies* 43, no. 3 (1997): 800–820.

Heider, Fritz, and Marianne Simmel, "An Experimental Study of Apparent Behavior." *American Journal of Psychology* 57 (1944): 23–41.

Heim, Michael. *The Metaphysics of Virtual Reality.* New York: Oxford University Press, 1993.

Herrnstein-Smith, Barbara. *Poetic Closure: A Study of How Poems End.* Chicago: University of Chicago Press, 1987.

Heyward, Michael. *The Ern Malley Affair.* London: Faber and Faber, 1993.

Higginson, Fred H., ed. *Anna Livia Plurabelle: The Making of a Chapter.* Minneapolis: University of Minnesota Press, 1960.

Hilts, Philip J. *Memory's Ghost: The Nature of Memory and the Strange Tale of Mr. M.* New York: Touchstone, 1995.

Hirsch, E. D, Jr. "Meaning and Significance Reinterpreted." *Critical Inquiry* 11 (1984): 202–25.

———. *Validity in Interpretation.* New Haven: Yale University Press, 1967.

Holub, Robert C. *Reception Theory: A Critical Introduction.* London: Methuen, 1984.

Homer. *The Odyssey.* Trans. Robert Fitzgerald. New York: Doubleday, 1961.

Hurtig, Brent. "The Plot Thickens." *New Media,* January 13, 1998, 36–42.

Huxley, Aldous. *Brave New World.* New York: Harper and Row, 1969.

Ingarden, Roman. *The Literary Work of Art: An Investigation on the Bor-*

derlines of Ontology, Logic, and Theory of Literature. Evanston, Ill.: Northwestern University Press.

Iser, Wolfgang. *The Act of Reading: A Theory of Aesthetic Response.* Baltimore: Johns Hopkins University Press, 1978.

———. "Interaction between Text and Reader." In *The Reader in the Text: Essays on Audience and Interpretation,* ed. Susan R. Suleiman and Inge Crosman. Princeton: Princeton University Press, 1980.

———. "The Reading Process: A Phenomenological Approach." In *Reader-Response Criticism: From Formalism to Post-Structuralism,* ed. Jane P. Tompkins. Baltimore: Johns Hopkins University Press, 1980.

James, Henry. *The Art of the Novel.* Ed. R. P. Blackmur. New York: Scribner's, 1934.

Johnson-Eilola, Johndan. "'Trying to See the Garden': Interdisciplinary Perspectives on Hypertext Use in Composition Instruction." *Writing on the Edge* 2, no. 2 (1991): 92–111.

Jones, Robert Alun, and Rand Spiro. "Imagined Conversations: The Relevance of Hypertext, Pragmatism, and Cognitive Flexibility Theory to the Interpretation of 'Classic Texts' in Intellectual History." Paper presented at the European Conference on Hypertext, Milan, December 1992.

Joyce, James. *Ulysses.* New York: Vintage, 1980.

Joyce, Michael. *afternoon, a story.* Storyspace software for Macintosh and Windows. Cambridge, Mass.: Eastgate Systems, 1990.

———. "Forms of Future." *Media-in-Transition,* November 5, 1997. <http://media-in-transition.mit.edu/articles/joyce.html>.

———. *Of Two Minds: Hypertext Pedagogy and Poetics.* Ann Arbor: University of Michigan Press, 1995.

———. "Ordinary Fiction." *Paradoxa* 4, no. 11 (spring 1999): 510–26.

———. "Siren Shapes: Exploratory and Constructive Hypertexts." *Academic Computing* 3, no. 4 (1988): 10–14, 37–42.

———. "Twelve Blue." *Post Modern Culture* 7, no. 3 (1997): <http://www.eastgate.com/TwelveBlue/>.

———. *WOE—or a Memory of What Will Be. Writing on the Edge* 2, no. 2 (1991). Storyspace software for Macintosh and Windows.

Katz, Alyssa. "So Many Words, So Few Megabytes." *Spin,* July 1994, 79.

Kelley, Sue. "A Literary Prize Gets Wired." *New Yorker,* September 14, 1998, 30–31.

Kermode, Frank. *The Sense of an Ending: Studies in the Theory of Narrative Fiction.* New York: Oxford University Press, 1966.

———. "Secrets and Narrative Sequence." In *On Narrative,* ed. W. J. T. Mitchell. Chicago: University of Chicago Press, 1981.

Kestner, Joseph. "The Novel and the Spatial Arts." In *Spatial Form in Narrative,* ed. Jeffrey R. Smitten and Ann Daghistany. Ithaca: Cornell University Press, 1981.

Kirchofer, Tom. "Hyperfiction Plot Moves at a Click." *AP Online,* March 14, 1998, <http://www.archives.nytimes.com/apoline.html>.

Landow, George P. *Hypertext: The Convergence of Contemporary Critical Theory and Technology.* Baltimore: Johns Hopkins University Press, 1992.

———. "Hypertext, Hypermedia, and Literary Studies: The State of the Art." In *Hypermedia and Literary Studies,* ed. Paul Delany and George P. Landow. Cambridge: MIT Press, 1991.

———. "Relationally Encoded Links and the Rhetoric of Hypertext." *Proceedings of Hypertext '87.* New York: Association of Computing Machinery, 1987.

———, ed. *The Dickens Web.* Storyspace hypertext software. Cambridge, Mass.: Eastgate Systems, 1992.

Landow, George P., and Paul Delany, eds. *Hypermedia and Literary Studies.* Cambridge: MIT Press, 1991.

Landow, George P., and Jon Lanestedt, eds. *The "In Memoriam" Web.* Storyspace hypertext software. Cambridge, Mass.: Eastgate Systems, 1993.

Lanham, Richard A. "The Electronic Word: Literary Study and the Digital Revolution." *New Literary History* 20 (1989): 265–89.

Larsen, Deena. *Marble Springs.* Storyspace hypertext software. Cambridge, Mass.: Eastgate Systems, 1994.

Latour, Bruno. "The Politics of Explanation: An Alternative." In *Knowledge and Reflexivity: New Frontiers in the Sociology of Knowledge,* ed. Steve Woolgar. London: Sage, 1988.

Leavis, Q. D. *Fiction and the Reading Public.* London: Chatto and Windus, 1932.

Leslie, Alan M., and Stephanie Keeble. "Do Six-Month-Old Infants Perceive Causality?" *Cognition* 25 (1987): 265–88.

Lillington, Karlin. "Now Read on, or Back, or Sideways, or Anywhere." *Guardian,* October 9, 1997, 1–3.

Lombreglia, Ralph. "The End of the Story." *Atlantic Unbound,* November 11, 1996. <http://www.theatlantic.com/unbound/digicult/dc9611/joyce.htm >.

Lyne, Adrian, dir. *Jacob's Ladder.* 1990.

Mailloux, Steven. *Interpretive Conventions: The Reader in the Study of American Fiction.* Ithaca: Cornell University Press, 1982.

Mayes, Terry, Michael R. Kibby, and Tony Anderson. "Signposts for Conceptual Orientation: Some Requirements for Learning from Hypertext." In *Hypertext: State of the Art,* ed. Ray McAleese. Oxford: Intellect Books, 1990.

McLuhan, Marshall. *The Gutenberg Galaxy: The Making of Typographic Man.* Toronto: University of Toronto Press, 1962.

———. *Understanding Media: The Extensions of Man.* New York: Signet, 1964.

Mechner, Jordan, dir. and prod. *The Last Express.* CD-ROM for Macintosh, Windows, and DOS. Novato, Calif.: Broderbund, 1997.

Mechner, Jordan, and Tomi Pierce. "The Making of *The Last Express:* An

Interview with Jordan Mechner and Tomi Pierce." Interview by Rick Barba. In *The Last Express: The Official Strategy Guide.* Rocklin, Calif.: Prima, 1997.

Meehan, James Richard. "The Metanovel: Writing Stories by Computer." Ph.D. diss., Yale University, 1976.

Melrod, George. "Digital Unbound." *Details,* October 1994, 162–65, 199.

Michotte, A. *The Perception of Causality.* New York: Basic Books, 1963.

Mickelsen, David. "Types of Spatial Structure in Narrative." In *Spatial Form in Narrative,* ed. Jeffrey R. Smitten and Ann Daghistany. Ithaca: Cornell University Press, 1981.

Miller, J. Hillis. *Fiction and Repetition: Seven English Novels.* Cambridge: Harvard University Press, 1982.

Miller, Laura. "www.claptrap.com." *New York Times Book Review,* March 15, 1998, 43.

Miller, Matthew. "Trip." *Postmodern Culture* 7, no. 1 (September 1996): <http://muse.jhu.edu/journals/postmodern_culture/v007/7.1miller.html>.

Miller, Robyn, and Rand Miller, dirs. and prods. *Myst.* CD-ROM for Macintosh. Novato, Calif.: Broderbund, 1993.

Mitchell, W. J. T. "Spatial Form in Literature: Toward a General Theory." In *The Language of Images,* ed. W. J. T. Mitchell. Chicago: University of Chicago Press, 1980.

———, ed. *On Narrative.* Chicago: University of Chicago Press, 1981.

McDaid, John. "Breaking Frames: Hyper-Mass Media." In *Hypertext/Hypermedia Handbook,* ed. Emily Berk and Joseph Devlin. New York: McGraw-Hill, 1991.

Mirapaul, Matthew. "A Vote of Confidence for Hypertext Fiction." *New York Times Cybertimes,* September 11, 1997, B10.

Monaco, James. *How to Read a Film: The Art, Technology, Language, History, and Theory of Film and Media.* New York: Oxford University Press, 1981.

Moulthrop, Stuart. "Beyond the Electronic Book: A Critique of Hypertext." In *Hypertext 1991 Proceedings,* ed. P. D. Stotts and R. K. Futura. New York: Association for Computing Machinery, 1991.

———. "Containing the Multitudes: The Problem of Closure in Interactive Fiction." *Association for Computers in the Humanities Newsletter* 10 (1988): 29–46.

———. "Forking Paths." Unpublished Storyspace software for Macintosh, 1986.

———. "Hypertext and 'the Hyperreal.'" *Hypertext '89 Proceedings,* 259–67. Pittsburgh: Association for Computing Machinery, 1989.

———. "In the Zones: Hypertext and the Politics of Interpretation." *Writing on the Edge* 1, no. 1 (1989): 18–27.

———. "The Politics of Hypertext." In *Evolving Perspectives on Computers and Composition Studies: Questions for the 1990's*, ed. Gail E. Hawisher and Cynthia L. Selfe. Urbana, Ill.: NCTE Press, 1991.

———. "Pushing Back: Living and Writing in Broken Space." *Modern Fiction Studies* 43, no. 3 (1997): 651–74.

———. "Reading from the Map: Metonymy and Metaphor in the Fiction of Forking Paths." In *Hypermedia and Literary Studies*, ed. Paul Delany and George P. Landow. Cambridge: MIT Press, 1991.

———. "Text, Authority, and the Fiction of Forking Paths." Typescript, 1987.

———. *Victory Garden.* Storyspace software for Macintosh and Windows. Cambridge, Mass.: Eastgate Systems, 1991.

Moulthrop, Stuart, and Nancy Kaplan. "Something to Imagine: Literature, Composition and Interactive Fiction." *Computers and Composition* 9, no. 1 (1991): 7–24.

———. "They Became What They Beheld: The Futility of Resistance in the Space of Hypertext Writing." In *Literacy and Computers: The Complications of Teaching and Learning with Technology*, ed. Susan J. Hilligoss and Cynthia L. Selfe. New York: Modern Language Association, 1991.

Murray, Janet H. *Hamlet on the Holodeck: The Future of Narrative in Cyberspace.* New York: Free Press, 1997.

Nell, Victor. *Lost in a Book: The Psychology of Reading for Pleasure.* New Haven: Yale University Press, 1988.

Nelson, Andrew, dir. and wr. *Titanic: Adventure out of Time.* CD-ROM for Macintosh and Windows. Knoxville: Cyberflix, 1996.

Nelson, Theodor Holm. *Literary Machines.* Guide hypertext software on disk for Macintosh. Bellevue, Wash.: OWL Systems, 1987.

Newsom, Gillian, and Eric Brown. "CD-ROM: What Went Wrong?" *New Media*, August 1998, 33–38.

Nielsen, Jakob. "The Art of Navigating through Hypertext." *Communications of the ACM* 33 (1990): 296–310.

———. *Hypertext and Hypermedia.* New York: Academic Press, 1990.

Niez, Anthony J., and Norman N. Holland. "Interactive Fiction." *Critical Inquiry* 11 (1984): 110–29.

Nyce, James M., and Paul Kahn, eds. *From Memex to Hypertext: Vannevar Bush and the Mind's Machine.* New York: Academic Press, 1992.

Olson, David R., and Nancy Torrance, eds. *Literacy and Orality.* New York: Cambridge University Press, 1991.

Ong, Walter J. *Orality and Literacy: The Technologizing of the Word.* London: Methuen, 1982.

Orasanu, Judith, ed. *Reading Comprehension: From Research to Practice.* Hillsdale, N.J.: Lawrence Erlbaum Associates, 1986.

Paul, Christiane. *Unreal City: A Hypertext Guide to T. S. Eliot's* The Waste Land. Storyspace software for Macintosh and Windows. Cambridge, Mass.: Eastgate Systems, 1995.

Paulson, William R. *The Noise of Culture: Literary Texts in a World of Information.* Ithaca: Cornell University Press, 1988.

Pinch, Trevor J., and Wiebe E. Bijker. "The Social Construction of Facts and Artifacts: or How the Sociology of Science and the Sociology of Technology Might Benefit Each Other." In *The Social Construction of Technological Systems: New Directions in the Sociology and History of Technology,* ed. Wiebe E. Bijker, Thomas P. Hughes, and Trevor J. Pinch. Cambridge: MIT Press, 1987.

Pinsky, Robert. "The Muse in the Machine: or, The Poetics of Zork." *New York Times Book Review,* March 19, 1995, 3, 26–27.

Plato. *Phaedrus.* Trans. Walter Hamilton. London: Penguin, 1973.

Pratt, Mary Louise. *Toward a Speech Act Theory of Literary Discourse.* Bloomington: Indiana University Press, 1977.

Richards, Neil. *Douglas Adams' Starship Titanic: The Official Strategy Guide.* New York: Three Rivers Press, 1998.

Ricoeur, Paul. *Interpretation Theory: Discourse and the Surplus of Meaning.* Fort Worth: Texas Christian University Press, 1976.

Riesman, David. *The Lonely Crowd: A Study of the Changing American Character.* New Haven: Yale University Press, 1950.

Rimmon-Kenan, Shlomith. *Narrative Fiction: Contemporary Poetics.* New York: Methuen, 1983.

Robbe-Grillet, Alain. *In the Labyrinth.* Trans. Christine Brooke-Rose. London: John Calder, 1980.

Rosenberg, Martin. "Contingency, Liberation, and the Seduction of Geometry: Hypertext as an Avant-garde Medium." *Perforations* 1, no. 3 (1992).

Rosenblatt, Louise M. *Literature as Exploration.* New York: Modern Language Association, 1968.

Rosenthal, Steve. "Douglas Engelbart." *Electric Word* 18 (1990): 21–25.

Ryan, Marie-Laure. *Possible Worlds, Artificial Intelligence, and Narrative Theory.* Bloomington: Indiana University Press, 1991.

Ryman, Geoff. *253.* <http://www.ryman-novel.com/info/home.htm>.

———. *253: The Print Remix.* London: Flamingo, 1998.

Samuelson, Pamela. "Some New Kinds of Authorship Made Possible by Computers and Some Intellectual Property Questions They Raise." *University of Pittsburgh Law Review* 53, no. 3 (1992): 586–704.

Sartre, Jean-Paul. *What Is Literature?* Trans. Bernard Frechtman. London: Philosophical Library, 1949.

Schank, Roger C. *Dynamic Memory: A Theory of Reminding and Learning in Computers and People.* Cambridge: Cambridge University Press, 1982.

Schank, Roger C., and Robert P. Abelson. *Scripts, Plans, Goals, and Understanding: An Inquiry into Human Knowledge Structures.* Hillsdale, N.J.: Lawrence Erlbaum Associates, 1977.

Scholes, Robert. *Textual Power: Literary Theory and the Teaching of English.* New Haven: Yale University Press, 1985.

Scholes, Robert, and Robert Kellogg. *The Nature of Narrative.* New York: Oxford University Press, 1966.

Shono, Haruhiko, dir. *Gadget: Invention, Travel, and Adventure.* CD-ROM. Synergy, 1994.

Slatin, John M. "Reading Hypertext: Order and Coherence in a New Medium." *College English,* 52, no. 8 (1990): 870–84.

Smith, Frank. *Understanding Reading: A Psycholinguistic Analysis of Reading and Learning to Read.* 3d ed. New York: Holt, Rinehart and Winston, 1982.

Smitten, Jeffrey R. "Spatial Form and Narrative Theory." In *Spatial Form in Narrative,* ed. Jeffrey R. Smitten and Ann Daghistany. Ithaca: Cornell University Press, 1981.

Smitten, Jeffrey R., and Ann Daghistany, eds. *Spatial Form in Narrative.* Ithaca: Cornell University Press, 1981.

Sperber, Deidre, and David Wilson. "Mutual Knowledge and Relevance in Theories of Comprehension." In *Mutual Knowledge,* ed. N. V. Smith. London: Academic Press, 1982.

Spiro, Rand J., Bertram C. Bruce, and William F. Brewer, eds. *Theoretical Issues in Reading Comprehension: Perspectives from Cognitive Psychology, Linguistics, Artificial Intelligence, and Education.* Hillsdale, N.J.: Lawrence Erlbaum Associates, 1980.

States, Bert O. *Dreaming and Storytelling.* Ithaca: Cornell University Press, 1993.

Stephenson, Wen. "Over the Edge: A Postmodern Freefall into Cyberspace." *Atlantic Unbound,* December 4, 1997, <http://www.theatlantic.com /unbound/media/ws9712.htm>.

Sterne, Laurence. *The Life and Opinions of Tristram Shandy.* 1759. London: Penguin Books, 1967.

Street, Brian V. *Literacy in Theory and Practice.* Cambridge: Cambridge University Press, 1984.

Swiss, Thomas. "Music and Noise: Marketing Hypertexts." Review of Eastgate Systems, Inc. *Post Modern Culture* 7, no. 1 (1996): <http:// jefferson.village.virginia.edu/pmc/text-only/issue.996/review-4.996>.

Tompkins, Jane, ed. *Reader-Response Criticism: From Formalism to Post-Structuralism.* Baltimore: Johns Hopkins University Press, 1980.

Trabasso, Tom, Tom Secco, and Paul Van Den Broek. "Causal Cohesion and Story Coherence." In *Learning and Comprehension of Text,* ed. Heinz Mandl, Nancy L. Stein, and Tom Trabasso. Hillsdale, N.J.: Lawrence Erlbaum Associates, 1984.

Trilling, Lionel. "Freud and Literature." In *The Liberal Imagination: Essays on Literature and Society.* New York: Viking, 1968.

Turkle, Sherry. *Life on the Screen: Identity in the Age of the Internet.* New York: Simon and Schuster, 1995.

Ulmer, Greg. "A Response to 'Twelve Blue' by Michael Joyce." *Postmodern*

Culture 8, no. 1 (1997): <http://jefferson.village.virginia.edu/pmc/issue.997/ulmer.997.html>.

van Dijk, Teun. *Macrostructures: An Interdisciplinary Study of Global Structures in Discourse, Interaction, and Cognition.* Hillsdale, N.J.: Lawrence Erlbaum Associates, 1980.

Vidan, Ivo. "Time Sequence in Spatial Fiction." In *Spatial Form in Narrative,* ed. Jeffrey R. Smitten and Ann Daghistany. Ithaca: Cornell University Press, 1981.

Waugh, Patricia. *Metafiction: The Theory and Practice of Self-Conscious Fiction.* London: Methuen, 1984.

Weschler, Lawrence. *Mr. Wilson's Cabinet of Wonder.* New York: Pantheon, 1995.

Wilson, Paul T., and Richard C. Anderson. "What They Don't Know Will Hurt Them: The Role of Prior Knowledge in Comprehension." In *Reading Comprehension: From Research to Practice,* ed. Judith Orasanu. Hillsdale, N.J.: Lawrence Erlbaum, 1986.

Wimsatt, W. K., and Monroe C. Beardsley. "The Intentional Fallacy." In *The Verbal Icon.* Lexington: University of Kentucky Press, 1954.

Wolfe, Tom. *The New Journalism.* London: Picador, 1972.

Wolff, Adam, Howard Cushnir, and Scott Kim. *Obsidian.* CD-ROM for Macintosh. Segasoft, 1996.

Woodhead, Nigel. *Hypertext and Hypermedia: Theory and Application.* Wilmslow, England: Sigma Press, 1991.

Woolf, Virginia. *Mrs Dalloway.* 1925. London: Grafton Books, 1976.

Wright, Patricia, and Ann Lickorish. "The Influence of Discourse Structure on Display and Navigation in Hypertexts." In *Computers and Writing: Models and Tools,* ed. Noel Williams and Patrik Holt. Oxford: Intellect Books, 1989.

Yankelovich, Nicole, Bernard J. Haan, Norman K. Meyrowitz, and Steven M. Drucker. "Intermedia: The Concept and Construction of a Seamless Information Environment." *Computer* 42 (1988): 81–96.

Ziegfeld, Richard. "Interactive Fiction: A New Literary Genre?" *New Literary History* 20 (1989): 340–73.

Index

Page numbers listed in italics refer to figures reproduced in the text.